abc's of Tape Recording

by Norman H. Crowhurst

HOWARD W. SAMS & CO., INC.
THE BOBBS-MERRILL CO., INC.
INDIANAPOLIS · KANSAS CITY · NEW YORK

THIRD EDITION

FIRST PRINTING—1971

Copyright © 1961, 1965, and 1971 by Howard W. Sams & Co., Inc., Indianapolis, Indiana 46206. Printed in the United States of America.

All rights reserved. Reproduction or use, without express permission, of editorial or pictorial content, in any manner, is prohibited. No patent liability is assumed with respect to the use of the information contained herein.

International Standard Book Number: 0-672-20805-9
Library of Congress Catalog Card Number: 78-135989

Preface

The possibility of recording sound—for entertainment as well as practical purposes—has always held a variety of fascinations. The advent of magnetic recording and the development of modern recording tape has brought this possibility within the means of all. Once you own a tape recorder, you'll find there are a multitude of uses for it—many you may have never dreamed of before.

To get the most from tape recording, you should know a little about how a recorder operates, how to choose a recorder, and how to get best results from it. This book covers all these points, plus many suggestions for other things you can do with recorders. Just as the automobile and telephone have become necessities, you'll wonder how you ever did without a tape recorder after you've had one a year or two.

You don't need a college degree to buy and use a tape recorder—nor to understand this book. Of necessity, the very fact that tape can be recorded on, played back, and erased for re-use means there are minor complications for avoiding mistakes in using a tape recorder. But a little care in getting to know your recorder—with the help of this book—will make operating it as simple as tuning a television receiver.

Have fun with your tape recorder!

NORMAN H. CROWHURST

Contents

CHAPTER 1

UNDERSTANDING TAPE RECORDERS 7
Transport Mechanism—Heads—Tracks and Alignment—Electronics—Controls

CHAPTER 2

CHOOSING A TAPE RECORDER 26
Price—What System?—Reel to Reel, Cartridge, or Cassette?—How Many Heads... Motors?—Push Buttons or Levers?—Other Controls—Judging Quality—Complete Recorder, or Deck and Electronics?—Supply Sources

CHAPTER 3

USING A TAPE RECORDER 51
Microphones—Recording Without a Microphone—Using the Right Level Setting—Tone Control—Multitrack Recorders—Care of Recorder and Tape

CHAPTER 4

SIMPLE THINGS TO DO WITH TAPE RECORDERS 68
Off-the-Air Recording—A Growing Family—Other Occasions—Portable Tape Recorders—Tape Letters

CHAPTER 5

PRACTICAL USES FOR TAPE RECORDERS 80
Using a Time Switch—Copying Tape—Self-Criticism—Party Games—Multiple Recordings—Learning a Language

CHAPTER 6

ADVANCED USES FOR TAPE RECORDERS 88
Automatic Message Taker — Time Indication — Home-Movie Sound—Transistor Microphone Amplifier

GLOSSARY 107

INDEX 110

1

Understanding

tape recorders

Whether you have already bought a tape recorder, or are only thinking of doing so, your first step toward getting maximum use from this wonderful invention is to understand it. Although the basic theory behind it involves the behavior of molecules in ways only partially understood by the most learned scientists, the use of these properties can be explained in a way that is very simple. And the number of possible uses is nearly infinite.

Every tape recorder has four main parts you'll need to know about: the transport mechanism, the heads, the electronics, and the controls. In each of these functions, various recorders will differ—but even the simplest must utilize all of them. The best or most expensive may have more or better features in these departments.

TRANSPORT MECHANISM

This is the main mechanical part of the recorder. Its job is to see that the tape passes the heads at a steady speed, both for making the recording and for playing it back. If the speed is not kept steady while you are either recording or playing back, the pitch of the sound you hear will go up and down and spoil the quality or effect. It may throw music off key, make voices sound too high or too low in pitch, or cause a warbling effect commonly called "wow" or a garbling effect called "flutter." A slow variation of speed causes wow; a fast variation causes flutter. If the speed is wrong but stays steady, the music will be off key or the pitch will be wrong.

Courtesy Kay Elemetrics Corp.

If you want to measure wow and flutter, you need an instrument such as this.

Several things are needed in a transport mechanism to ensure that the speed remains steady (Fig. 1-1). One is a steady drive to pull the tape past the heads. This requires a motor, either electric or spring-wound, that is exceptionally free-moving and steady in its torque, or pull. To help hold this movement steady, there is usually a carefully balanced flywheel attached to the drum or cylinder that moves the tape, called a capstan. As it rotates, its weight helps steady the movement.

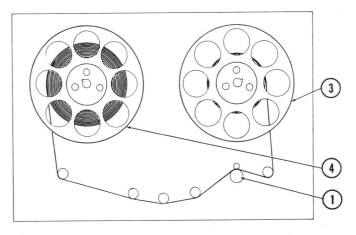

① DRIVE MUST BE SMOOTH AND CONSTANT.

② TAPE MUST MOVE SMOOTHLY PAST ALL GUIDES AND HEADS, WITHOUT SNATCHING OR CHATTERING.

③ PULL OF TAKE-UP REEL MUST BE GENTLE AND SMOOTH.

④ DRAG OF FEED REEL MUST BE LIGHT AND SMOOTH.

⑤ PATH OF TAPE MUST PROVIDE STEADY UNWOBBLING MOTION.

⑥ PRESSURE OF TAPE AGAINST HEADS MUST BE FLAT, UNIFORM, AND STEADY (WHETHER OR NOT PRESSURE PADS ARE USED).

⑦ GUIDES FOR TAPE MUST ENSURE ACCURATE LOCATION OF PATH BY HEADS, IN CORRECT POSITION, AND IN CORRECT ALIGNMENT.

Fig. 1-1. Main features to watch for in a tape transport mechanism; the actual layout may vary, but here are the important properties each should have.

Finally, the tape must be handled very smoothly as it goes to and comes from the heads. If the guide posts, or the spools that feed or take up the tape, produce any "snatching," the movement will be jerky instead of smooth.

In the older tape systems, spools are used to store the tape, and also to feed it into the machine and rewind it after playing. These spools are really part of the transport mechanism. In more modern machines, the tape is stored in cartridges; the spool is inside the cartridge case, and this whole package becomes part of the transport while the tape is playing.

The original RCA tape cartridge contained both spools as does the newer Philips *cassette*. In the CBS/3M cartridge, however, only the feed spool is in the cartridge. The take-up spool (which winds up the tape after it has passed the head mechanism) is a permanent part of the machine. The same spool is used for every tape played.

Another series of popular cartridges, standardized by NAB (National Association of Broadcasters), uses an endless piece of tape (the ends are joined together to make an endless loop). This cartridge is played by withdrawing the tape from the center of the spool and passing it through the transport mechanism which projects into an appropriate receptor on the cartridge. Then the tape is fed back onto the outside of the same spool.

Several speeds are used, for different purposes. The older machines provide 15, 7½, or 3¾ inches per second (ips). Some have provision for changing gears, belts, or pulleys to provide more than one speed. A few machines have a continuously variable speed adjustment instead of just two or three fixed speeds. Although the continuous adjustment is an attractive refinement for some purposes, most recorders incorporating it are less able to hold a steady speed. So, for most purposes, the fixed speeds are better—which is why more machines are built that way.

Dictation machines run at 3¾ or 1⅞ ips, and their frequency response is inferior to that of professional or music-quality machines. The newer machines for playing cartridges use the speeds the cartridges are designed to play: 3¾ ips for the RCA type, and 1⅞ ips for the Philips cassette and the CBS/3M types. The NAB standard single-spool cartridges use tape recorded at any of the previous standard speeds of 7½, 3¾, or 1⅞ ips.

In most machines, the transport mechanism makes the tape travel in only one direction—either left to right or right to left, according to the design of the machine. If there is only one track on the tape (two for stereo), this is all you need. But where a tape has two or more tracks that are recorded in opposite directions, the machine must be able to move the tape both ways.

Most machines require that you put the tape through the machine again, by interchanging the feed and take-up spools. But a few provide for automatic reversal of the tape travel in the transport mechanism (and a corresponding change in position of the heads), so the correct tracks on the tape can be used in either direction. This can be a great convenience, because the tape direction can be reversed with almost no time lost. But there is one disadvantage—accurate head alignment for recording and playing both ways is much more difficult to achieve. Some cartridge machines have provision for such automatic reversal at the end of the first track.

Another difference in transport mechanisms is in the number of motors used. The simplest machines use one motor to drive the capstan and the take-up spool. The same motor is used, through an arrangement of clutches and pulleys, to move the tape forward and also rewind it.

Using one motor for several purposes imposes an extra load that can make its motion jerky. To avoid this difficulty, many of the better machines use two or even three motors. One motor drives the capstan and another motor (or two) the take-up and rewind spools. The change from the gentle pull needed to wind the tape, to the stronger pull needed for fast forward or rewind, is achieved either by a mechanical clutch and pulley, or by electrical switching in the motor circuit.

Another important property of a transport mechanism is its ability to let you quickly control the tape while hunting back and forth for one particular spot. Especially important is the way the brakes must work to stop the tape from its high speed. The brakes may be mechanical, like those used on car wheels, or electrical like those used on some electric trains, where a different way of connecting the motor turns it into a brake. What is more important than the type of brake is how well it does its job. The best way to find out how well any brake works is to give it a whirl!

HEADS

In every tape recorder except those designed for playing only, the heads perform three functions—record, playback, and erase. The better machines have three heads, one for each function. This is the preferred arrangement, because for optimum performance the heads should be built a little differently from each other. But a great many recorders are able to give quite acceptable quality with only two heads by combining the record and playback functions into one head. Of course, machines intended solely for playing tapes need only a playback head.

To understand the different heads, we need to know how magnetism is impressed onto the tape. Whenever current flows through a coil, its core becomes magnetized. If the current is made to alternate by an audio signal,

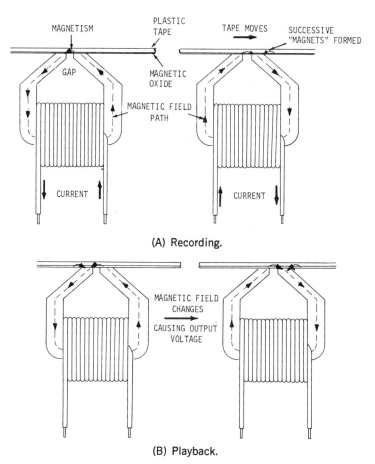

(A) Recording.

(B) Playback.

Fig. 1-2. How magnetic recording works: (A) Currents in the record head produce tiny magnets on the surface of the tape. (B) Passage of these magnets past the playback head produces an output voltage, which is then amplified.

the core will experience corresponding magnetic alternations. In a tape recorder, the magnetic tape takes the place of a very small piece of such a core. As it passes the gap, the tape is magnetized in accordance with the audio current at that instant (Fig. 1-2).

When a tape is played back, the "magnets" that were put onto the tape during record are successively placed in the gap of the playback-head core. The changing magnetism produces voltage fluctuations in the coil that is wound around the core. These voltage fluctuations are then amplified and reproduced as a replica of the original sound.

The tape must pass the record and playback heads at a constant speed, to ensure exact correspondence of the rate of audible fluctuations. This means that a very high audio frequency, corresponding to a shrill musical pitch or sound, will produce extremely short magnets on the tape; conversely a low audio frequency, or bass note, will produce comparatively long magnets.

Suppose the tape travels at $3\frac{3}{4}$ ips and the lowest frequency is 40 vibrations a second. This amounts to slightly more than 10 vibrations for each inch of tape. Since each vibration is a "to" and a "fro," it will correspond to two magnets, one in each direction along the tape. Hence, each magnet will be less than one-twentieth of an inch long.

That doesn't sound long does it—one twentieth of an inch? But compare it with the 1/8000-inch magnets found at 16,000 vibrations per second (the highest frequency audible to most humans). Now the tape must accommodate 32,000 magnetic alternations in roughly four inches of tape, meaning each magnet is 1/8000 inch. Considering that this is a "short" magnet, we can understand why a 1/20-inch magnet is considered a long one.

The fact that such extremely short magnets are impressed on the tape means that the gap in the core of the head must be extremely narrow—not more than 1/8000 inch in the example just mentioned. The head gap must be narrow enough to handle the highest audio frequency required at the normal speed of the tape. In other words, the slower the tape speed, the narrower the head gap must be.

Erase

There are two types of erase heads. The simpler one is merely a permanent magnet that is brought into contact with the tape when the erase function is required. For many purposes this is quite adequate. However, the permanent magnet leaves the tape magnetized in one direction as it goes away from the erase head. A background hiss or rumble results, because of fluctuations in contact pressure between the tape and the magnet as the tape passes. This background noise cannot be entirely eliminated, but for many purposes it is not bothersome, unless you are interested in recording symphonic music, for instance.

The better erase heads use current of ultrasonic frequency, usually between 50,000 and 100,000 Hz. This is

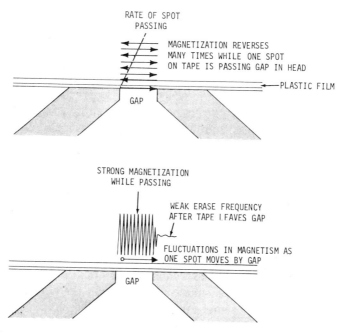

Fig. 1-3. How high-frequency erase works: (Top) While one spot is passing the gap, the magnetization across the gap will change many times. (Bottom) As the spot leaves the gap, the intensity of high-frequency magnetization diminishes, leaving very little in the tape, and nothing of any lower frequency.

above the range of human hearing; hence, any background noise on the tape will not be audible. Of the three types, the erase head has the widest magnetic gap against the tape. This means the current undergoes many reversals while one particle of the tape is passing the gap. In consequence, the high-frequency alternations of magnetism in the tape have time to die away quite gradually as the tape is leaving the gap, and the tape comes away almost completely demagnetized. What little is left is alternating at this ultrasonic frequency and thus is not audible (Fig. 1-3).

A good ultrasonic erase-current supply, which means the alternations must be equal in both directions and also very smooth, together with an erase head that maintains proper contact with the tape, results in complete erasure of the tape. This gives a noiseless background, so that any new material will not be marred by traces of the old recording on the track.

Record

In machines that have separate record and playback heads, the magnetic gap of the record head is intermediate in width between the erase and playback heads. If the record head used only currents corresponding to the audible frequencies to be recorded, its gap would need to be as narrow as that of the playback head. But direct recording of program currents in this way results in quite a large amount of distortion because of the magnetic characteristics of the tape, as explained earlier. Different magnitudes of current do not produce equivalent changes in magnetization; consequently, the sound waveform becomes distorted.

To avoid this distortion, the program current is normally mixed with current having an ultrasonic frequency. When the magnitude of this ultrasonic current (called *bias current*) is correctly adjusted, any distortion due to the magnetic characteristics will virtually disappear. All but the cheapest tape recorders use this form of high-frequency bias.

A record head with high-frequency bias works much like the ultrasonic erase head: Many alternations of the ultrasonic frequency occur as the tape passes the head. This high frequency dies out in the tape, leaving only the lower-frequency audible program material on the tape as it goes past the magnetic gap of the record head (Fig. 1-4).

Because of the accompanying ultrasonic bias, the gap can be made wider than the length of the individual "magnets" corresponding to the highest audio frequency put onto the tape. In fact it normally is, to avoid distortion of the lowest audio frequencies to be recorded and to achieve good penetration of the magnetism into the tape. Without ultrasonic bias, the gap would have to be

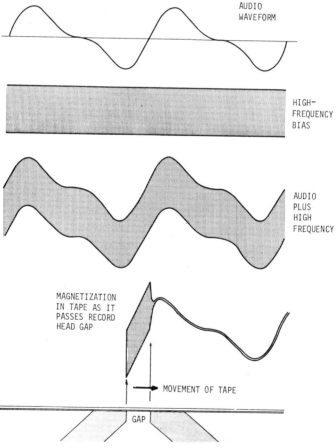

Fig. 1-4. How high-frequency bias works.

narrow enough to impress the highest audio frequencies on the tape as it passes.

Cross-Field Heads

The idea of using interacting heads is not new, but has only recently found its way into practical use. Two different methods should be distinguished. In the type first embodied in certain models by Roberts, the heads are located on opposite sides of the tape (Fig. 1-5). The audio is fed to the head on the coated side of the tape, while the bias is fed to a much more widely gapped head on the opposite side of the tape.

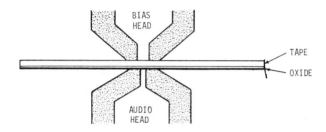

Fig. 1-5. An arrangement to produce cross-field magnetization.

In the other method, the function of the additional head, which is part of the same composite assembly, is slightly different (Fig. 1-6). By applying the bias to a different gap located on the same (coated) side of the tape as the audio gap, the contour of the magnetic field through which the tape progresses is changed. This contour produces a deeper penetration of the recorded magnetization into the tape, especially at the important higher frequencies, and a better solution of the problem, previously encountered, of achieving good high-frequency response and level with low distortion. Previously, adjustment of bias would require one setting for best high-frequency performance, another for maximum recorded level, and yet a third for minimum distortion at lower frequencies. Providing the separate gaps that contour the magnetic field makes it possible to achieve better results in all three respects at the same time.

17

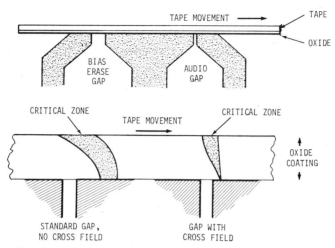

Fig. 1-6. Cross-field magnetization using a single head assembly.

Playback

The playback head must have a gap narrow enough to handle the tiny magnets corresponding to the highest frequencies recorded on the tape. This is why the best machines have separate record and playback heads. By using a recording head with a larger gap, it is easier to give the tape a high magnetic intensity; then the narrowest possible gap on the playback head makes it possible to take off more of the highest frequencies recorded on it.

Medium- and lower-priced recorders use a compromise head that combines the functions of record and playback. The same head is fed a current to magnetize the tape during record, and is made to pick up the voltages to be amplified during playback. The head gap is a compromise width that puts a reasonably good magnetic intensity onto the tape during record and picks off a reasonably high frequency during playback. A certain amount of quality is sacrified at both ends by making this compromise, but the results are good enough for most purposes.

Another advantage of using separate heads is that you can record and play back at once. While you are recording, the separate playback head allows you, at the same instant, to hear exactly what you are recording. Thus, you immediately know when something goes wrong, without wasting a whole tape or perhaps missing a chance to record a once-in-a-lifetime event.

TRACKS AND ALIGNMENT

Heads are quite critical in several ways. As well as needing the right width (which is measured along the direction the tape travels), the gap must have the right overall dimension and be correctly positioned, especially with tapes that carry more than one track (Fig. 1-7). The width of the core must match the width of the track on the tape. For full-track tapes, this is almost one-fourth of an inch. For two-track tapes (which use either one in each direction, or two in the same direction for stereo), each track is less than one-tenth of an inch wide and the core must have a corresponding width. For four-track tapes, allowing four ordinary or two stereo sides, each track is about one-thirtieth of an inch wide. The tracks on the CBS/3M system are about the same width. So the core of each head (one for each track) must have the same width.

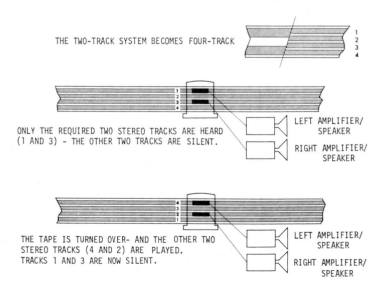

Courtesy Magnetic Recording Industry Association

Fig. 1-7. Placement of heads and tracks on two- and four-track tapes.

That was the first step from single and double track. Next, the track width and spacing between was halved again, to squeeze eight tracks onto ¼-inch tape. This was made possible only by improvement in the properties of the magnetic material on the tape.

In the early days of two-track stereo, two kinds of head placement were used: in-line and staggered. The problem was that of securing adequate screening between playback heads, so that crosstalk did not occur between tracks. Staggering overcame that, but introduced a more difficult problem: timing.

If the heads are staggered, the same "time" in the program occurs at different points along the tape. Splicing is virtually impossible. But also, if the speed is off, which naturally makes the pitch off too, timing goes wrong. There will be a time difference between the program on both tracks that will destroy the correct stereo illusion. Exactly correct speed is necessary, not only for precise pitch, but for stereo synchronism.

In-line heads, where both heads pick up sound at exactly the same point along the tape, avoid this problem completely, and splicing is no problem either. For four-track and eight-track tapes, the same is true. But there is another way that heads can be staggered. Pairs of heads that play the same stereo program must always be in-line. However, there is no reason why heads that play different programs should be in-line.

If the same heads play all the tracks, by a lateral shift, this question does not arise. And as a matter of economy, most multitrack recorders do use the same pair of heads for each stereo pair of tracks, merely by shifting them appropriately.

However, the economy question is not that simple in favor of only two heads. Moving the heads across the tape, to locate at precisely the correct positions for each pair of tracks, involves precision movement, and the mechanism to achieve it reliably. On the other hand, if the heads are permanently installed for each track, they can be independently and permanently adjusted for position and alignment, and track changing is then achieved by switching head outputs.

With the advent of cartridges and cassettes, and of wider tape, such as ½-inch or 1-inch for high-quality master or submaster tapes, the number of tracks on the tape increases the variety of track widths involved. Having the head the right shape and size for the tracks on the tape is only part of the problem. The rest is concerned with having the head properly aligned with the tracks on the tape. Correct alignment requires the head to be centered on the tracks, and the gap to be precisely at right angles to the tape movement. Of the two, the latter is definitely more important. The slightest twist in either the mounting of the head or the passage of the tape past it (Fig.

(A) Head mounted crooked.   (B) Tape twists while passing head.

Fig. 1-8. Correct alignment can fail because the head mounting is crooked or the tape does not stay straight.

1-8) will result in drastic loss of the higher frequencies, because the gap is so narrow that it will no longer line up with the tiny magnets on the tape.

ELECTRONICS

In this book, we are not going to explain amplification in detail, nor go into equalization. But some persons have been known to buy a tape *deck*, under the impression they were buying a tape recorder, only to find it would not work without adding some electronic accessories.

It's true some high-fidelity preamps are now providing a tape-recorder input position so the tape deck does not need its own electronics for playback *if the two are properly matched*. For this, you need to check the playback head mounted on the deck to make sure its impedance suits the input provided by the preamp. Even then, you will need additional electronics for recording, if you want to make your own tapes. This includes an audio feed to drive the record head, as well as a bias-and-erase oscillator.

A two-head machine (erase and record/playback) will usually have a bias oscillator, plus an amplifier that serves the dual purpose of record and playback. A few of the very cheap recorders do not have a bias oscillator—and the quality of sound recorded by them shows it.

In the record position, the amplifier input is connected to the microphone socket or to the phono-radio-TV input, and the output is combined with high-frequency currents from the bias oscillator and fed to the head to make a recording.

In the playback position, the head is connected to the amplifier input and the output to a speaker, an output socket, or both. Often the output socket is a jack arranged so that when you insert the corresponding plug, the internal speaker is disconnected. In the record position, the output may be available for listening and for driving the tape head; but it usually can be disconnected from the built-in speaker, as a precaution against howl due to acoustic feedback. But when you insert the plug into the jack, the disconnecting function of the switch may be bypassed, so you can listen on headphones to what you are recording.

Listening on headphones is not quite as good as using a separate playback head to monitor what you record. For example, if the head should be dirty, the recording quality would be poor. But you would not know this until you had rewound the tape and played it back, because you are listening to what is being recorded, not to what you have actually recorded.

This is the main economy of the two-head recorder—it uses only one amplifier for both record and playback (two for stereo, one per channel).

Three-head machines not only have separate amplifiers for each channel, but also for record and playback. This makes it possible to monitor your recordings, as well as to perform a number of stunts, as you'll see in later chapters. But your main reason for choosing a three-head machine will probably be its superior quality of sound, especially if you are interested in truly high-fidelity recording.

Incorporated into tape amplifiers is a form of electronic circuit action called equalization. From the explanation of how tapes are made and reproduced, you will recall that the strength of the magnets made on the tape depends on the amount of current through the record head. Likewise, the strength of the output voltage depends on the speed at which the magnets pass the playback head. (This principle is not affected if the same head doubles for both playback and record.) The longer, low-frequency magnets require more time to pass the small head gap than do the much shorter, high-frequency magnets. So the output voltage is much smaller at low than at high frequencies.

Equalization corrects this difference, partially during record, by emphasizing the low frequencies a little more than the high. The remainder is corrected during playback, when more boost is given to the low frequencies. In other words, the function of equalization is to make sure all frequencies in the original sound are reproduced correctly.

When you buy prerecorded tapes, the equalization provided in the playback position of your recorder should be correct for that used for the prerecorded tapes. Equalization has been standardized by the NAB (National Association of Broadcasters) and the MRIA (Magnetic Recording Industry Association). Unless your machine has this standard equalization, you will not get the proper results.

Some recorders do not have this standard equalization. If you don't plan to play prerecorded tapes, these recorders may produce adequate quality for your needs.

The bias oscillator can also affect the quality. To give good quality, its output current needs to have a smooth, balanced waveform. If not, the quality of sound recorded will be distorted. Recordings made in another machine will not suffer, because the bias oscillator is idle during playback.

The bias current also needs to be adjusted to the right intensity for the tape used in the machine. Too much bias current usually results in loss of the high frequencies recorded; too little results in more distortion, particularly of the lower frequencies. This adjustment is something you should not undertake without proper equipment. So if you suspect the bias current is wrong, you should have it corrected by a service technician.

CONTROLS

Even the simplest recorder must have some controls, while the more flexible types, capable of doing most of the interesting things we shall describe presently, must have quite a lot of controls. Ever since the first tape recorders were made, designers have had the problem of making the controls simple to operate, yet flexible enough that the user can make the recorder a highly useful and entertaining medium.

Quite a lot of attention has also gone into making the controls as foolproof as possible—so you won't find afterward that you never recorded a word because of a silly mistake, or that you erased your favorite recording unintentionally because you forgot to switch something on or off.

The latter is more of a problem. In most recorders, anything already on the tape is erased when the machine is switched to record. Various means are employed to make sure you do not place the recorder in the record position by mistake. Many recorders have a safety button or lever which you must hold down while pressing the lever or button that moves the tape. When you stop it, the safety lever or button automatically resets; and you have to go through the same contortions to make it start recording again. Some recorders merely rely on a red light that glows to warn you whenever the machine is in the record position.

Many machines have interlocks between the fast forward and rewind positions. These prevent you from switching to the playback or record position unless you have first stopped the tape. Otherwise, the tape may spill all over the place or may break—two misfortunes that can very easily happen without such a safeguard.

The controls can be grouped into two headings. The mechanical controls start and stop the transport mechanism, bring the tape into contact with the heads, or run it in fast forward or reverse. The electrical controls select the record or playback function (or select amplifiers in the three-head machines), adjust the volume for record or playback, and sometimes provide for adjustment of the tonal quality.

For convenience, the two controls are combined for some functions. For example, the selection of record or playback is often combined with the transport control.

Levers and push buttons are the two main types of mechanical control. With push buttons, every action is controlled by pushing a different button, and interlock is attained by the locking of buttons that must not operate. For example, whenever the fast forward or rewind button is down, the record or playback button will be locked. These can be operated only *after* the stop button has been pushed.

When levers are used, there may be several with some kind of interlock that prevents the record or playback lever from being moved unless the tape has been stopped. Sometimes just one lever is used that moves in two or three directions and provides all functions. Interlock is achieved by the simple method of having the lever go through the stop position each time it must be moved to a new one.

Whether to buy a push-button recorder or one with levers is a matter of personal preference, although to some extent it depends on what you plan to do with the recorder. Some uses are better handled with a push-button control system; with others, a lever control is more convenient. These details will be discussed in the following chapter.

2

Choosing a

tape recorder

Armed with the understanding obtained from the previous chapter, you are in a much better position to choose a tape recorder that will suit your needs. First you should make up your mind what you want to use it for, and then how much you are prepared to pay.

PRICE

Although you can usually get a better recorder by spending more money, a high price is not always a guarantee of better performance; nor does it mean the recorder is best suited to your needs. A high-quality professional machine would be a waste of money if used only for dictation, and a dictating machine is no good for recording symphony music. Yet each is designed to do its own job well.

A tape machine is both a mechanical and an electrical device. Some manufacturers emphasize the one more than the other. To some extent, quality depends on price; but in most price groups (except possibly the highest), the quality will deviate in one of these two areas.

If you are concerned mainly with providing acceptable music for dancing, a good transport mechanism will ensure good tempo and freedom from wow and flutter. On the other hand, the amount of distortion may not be acceptable for rendering concert music.

There is less to be said for a machine with first-class electronics and not-so-good mechanical design. If the transport is deficient, no degree of quality in electronics will make the machine sound good.

WHAT SYSTEM?

The first main area of confusion to settle in your mind is: "What system do I want to use? How many tracks, and what speeds?" If you want to make your own tapes, it's best to stick with the established sizes and speeds. Prerecorded tapes will continue to be made available in these sizes and speeds, even after newer systems gain hold.

If you're interested in stereo also, get a machine that will handle either two- or four-track tapes at both the 3¾- and 7½-inch speeds. Because of the placement of the pairs of tracks used in each direction for stereo on a four-track tape, the same heads can be used for two-track tape, just by positioning them over the two tracks.

Now that most machines are available in stereo models, you would do well to consider this possibility, even if you are not immediately interested in recording your own stereo. The additional facilities needed for stereo can also be used for adding echo effects, making multiple recordings, providing self-criticism facilities in learning a language or improving your public speaking voice, and many others. These are explained later in the book when the various things you can do with a tape recorder are discussed.

REEL TO REEL, CARTRIDGE, OR CASSETTE?

When the first edition of this book was printed, all tape recorders, except for some experimental types, were reel to reel. The position is rapidly reversing, so that reel to reel will presently become almost a museum piece. However, it is the opinion of people in the industry that reel to reel will always have a market, because of the high-quality systems already in the field, and because it has certain flexibilities not so readily available in cartridge or cassette.

When cartridge and cassette recorders first appeared on the scene, they were strictly a "play only" proposition for playing prerecorded tapes. But now the majority of cartridge and cassette recorders at least have a model that provides for recording as well as playing. It may even be that "play only" machines are already in a minority.

The big advantage to cartridge or cassette is the ease of loading. Reel to reel always presents the problems of spilling tape and incorrect threading. Using reel to reel is never as easy as putting a disc on a phonograph. The advent of progressively more reliable cartridges and cassettes has almost reversed this situation.

Now tape is at least as easy to load as the phonograph disc, if not easier. And it does not have the problem that phonographs do of susceptibility to mechanical vibration. However well a phonograph is built, the needle can jump grooves when the phonograph is subjected to vibration. For example, heavy dancing on a floor that transmits vibration to the phonograph will cause the needle to jump grooves.

Careful adjustment can eliminate this, but the adjustment may need maintaining. It involves a fine balance between dynamic and gravity balance. Slight deviation from the perfection of balance will quickly render the phonograph susceptible to groove-jumping triggered by mechanical vibrations. Tape recorders are not susceptible to this at all.

Cassettes and cartridges are easy to insert into the slot provided on the machine. There are three basic types of prepackage in which tape can thus be supplied, and these are not interchangeable. Each must be used with a machine designed to receive that type of tape container, or vice versa.

The first type of prepackage tape to appear is now generally called "cassette," although it comes in a variety of detail designs. These designs are not interchangeable unless the exterior fittings are identical, in which case internal differences are unimportant. The first to appear was the RCA cassette, but the smaller ones, like those developed by Philips (Norelco), have become the most popular.

This type is characterized by having two spools of tape; it plays half its tracks in one direction and half in the other direction. Essentially, in this type, there is a reel-to-reel arrangement in a package that can readily be lifted from the recorder and replaced with an identical package in a matter of seconds.

Playing the recorded material on tracks going in opposite directions may be achieved, in the simplest machines, by flipping the cassette: Put it in one way to play tracks going one way, then flip it and reinsert it to play the tracks playing back again. Philips (Norelco) markets an attachment that will play a number of these cassettes first on one side, flipping each in turn, ready to play the other side.

Inserting a number of cassettes into a machine with this attachment will cause it to play all the No. 1 sides in turn, then all the No. 2 sides in turn, then all the No. 1 sides in turn again, indefinitely. Consequently, a long, continuous program can be arranged.

More complicated machines reverse the tape automatically when the end is reached, and have heads that will play each track in turn, either by moving laterally or by having different sets of heads that are switched. Each kind of machine needs a device to let the machine "know" when a "side" is finished, to initiate the mechanism that performs the next step: changing the cassette, or reversing the direction. The same device on the cartridge, using the tape itself as it nears the end of the reel, serves either purpose.

Both the other kinds of cartridge use only one spool of tape. In the first to appear on the market, developed by CBS/3M, each cartridge contains a single spool of tape, with one end free. The machine takes this end, passes it

TYPICAL STEREO TAPE RECORDERS

Norelco Model 4408.

Concertone Model 790.

V-M Model 748.

Ferrograph Series Seven.

Roberts Model 1730.

Lafayette Model RK-960.

Wollensak 6100 Series.

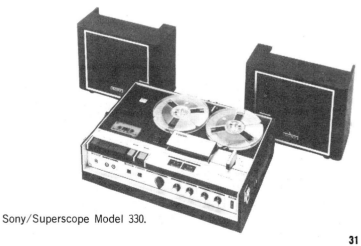

Sony/Superscope Model 330.

through the playing mechanism onto a winding spool within the machine, plays the tape through, then rewinds it back onto the cartridge spool and moves the cartridge away to bring another one into position for playing.

The one that is achieving far wider use employs an endless spool of tape: The cartridge is filled with tape, and the two ends are spliced together. Tape is withdrawn from the center of the cartridge, fed through the playing mechanism (or recording mechanism) and back onto the outside of the spool. As the tape plays, moving inward through successive layers, it moves forward at an increasing turn rate, but at the same linear (inches per second) rate, thus causing slight slippage between adjacent turns of tape. This means that the design of the cartridge must provide for feeding it correctly, when the right amount of tape is put in. Too much would prevent the slippage freely enough, and cause binding. Too little would not retain proper control of tape movement.

HOW MANY HEADS . . . MOTORS?

Does your choice rest between a two-head or a three-head machine? There is quite a price differential, because the latter has an extra head and almost twice the amount of electronics. This makes a strong argument for the two-head machine, unless you really are interested in making professional-quality recordings. Another point favoring a three-head stereo machine is its greater flexibility for some of the more advanced uses—it's almost as good as having two machines.

Do you need two or three motors, or is one enough? Here again, the extra motors add to the cost, so you need to weigh this extra cost against what you get. Some of the single-motor machines perform very well in spite of the design limitation imposed. When making your decision, allow for the fact that mechanical performance can deteriorate in time. The two- or three-motor machines are likely to retain a constant speed almost indefinitely, because the capstan motor runs under practically no load. Where one motor has to pull the take-up reel and provide fast forward and rewind action, there is more likelihood of wear and consequent deterioration. In the two- or three-motor machines, the speed constancy depends on the capstan motor; wear in the motors used for take-up and fast forward or rewind will not affect it.

On the other side of the choice factor—if you have a budget to consider—one good motor is better than two or three cheap ones. Motors for driving tape recorders must be precision-built—in fact, many manufacturers buy new stock-production motors and rebuild them, because they are not quite good enough as bought.

The real test is how much wow or flutter you get. The best way to check, aside from measurement, is by listening to piano recordings. No musical instrument shows up wow or flutter as readily as a piano, because it produces no mechanical tremolo or vibrato effect. Practically all other instruments can do so at the discretion of the player, even an organ. For this reason, speed variations can be more readily detected in recorded piano music.

PUSH BUTTONS OR LEVERS?

Whether to buy a machine that has push buttons, or one that has levers, is a matter of personal choice, based to some extent on how you plan to use the machine. A well-designed single-lever arrangement probably gives the fastest and surest control of the tape when you want to find an exact spot on it, do precise editing, and so on. But where such precision control is not required, the push buttons are more convenient. If you are just a casual user, you'll find it much easier to hit a button than to shift a lever!

Either type of control can do its job well—or not so well. So, how well the controls do their job may prove more important than which type they are. Take five or ten minutes attempting to master the control of any machine you are considering buying. If within that time it doesn't always do what you want it to do, look around for another machine. Of course, you'll need a little time to become familiar with a new mechanism. If it is a good mechanism, however, it shouldn't take you too long to find that out.

EXAMPLES OF PROFESSIONAL-TYPE EQUIPMENT USING SEPARATE DECK AND ELECTRONICS

Sony/Superscope Model 850.

Crown 800 Series.

Ampex Model 351-C.

Viking Model 433W.

Pioneer Model T-600.

Crown 700 Series.

Wollensak 5700 Series.

Roberts Model 5050XD.

Allied Model TD-1099.

This brings us to the other controls. For example, what arrangements (if any) are available for foolproof protection against accidental erasure? The question of whether a light is enough, or whether you need a mechanical interlock, to safeguard you against making a mistake depends largely on you. If you're accustomed to a machine with a safety interlock, don't ever try to do without one. It will only be a matter of time before you erase something from the tape you didn't want to.

A useful feature on some recorders is a pause button. It differs from the stop button in that the recorder remains at a standstill only while the button is held down—none of the other settings are changed. As soon as the pause button is released, recording continues—there is no need to depress the record button again. At the same time, you cannot forget and leave the machine in record, because the record setting is canceled when the stop button is depressed.

If a tape recorder is new to you, you may be able to discipline yourself to watch for the red light at all times. Even then, it is always safer to take the machine out of record every time you stop it. It's so easy, when you are recording intermittently, to leave the machine on record "to save trouble." Before you know it, you'll end up running the tape again while it's still in record. This is just what the safety interlock is intended to prevent.

Should the machine be arranged to play the tape in both directions, or in only one? One make of machine plays tape in both directions merely by the turn of a switch. This feature has held customers to that one make for many years—to them, this one advantage is worth more than any others its competitors might have. The main problem of these machines is maintaining proper head alignment. For this reason, many are operating far below their capabilities.

If quick reversal is inviting to you—say you want several hours of uninterrupted programming—then you may prefer this feature in spite of the possible sacrifice in quality. But if one side of the tape will carry all you want most of the time, then you will most likely prefer another make.

OTHER CONTROLS

Some machines (notably the three-head types) have separate volume controls for record and playback; others have one that combines both functions. Actually, the number of controls is not as important as whether the record and playback levels are correctly set. A single control will have to be readjusted every time you switch from record to playback or vice versa, whereas separate controls can be left where they are.

That last statement would appear to favor separate controls. But the advantage is not as great as you might expect. To get the best from your recorder, you must always be conscious of the volume setting. Separate controls may lull you into forgetting they exist; hence, you could lose quality by recording at the wrong level.

JUDGING QUALITY

There is considerable difference in quality between the performance of different recorders. So, before making a choice, you need to know how to judge good quality.

We've already mentioned that a piano recording will show up wow and flutter better than any other musical instrument will. Freedom from individual forms of distortion can best be judged by listening to a recording of a small combo that has just a few, well-separated musical instruments including good bass, clear, round middle tones, and some good, clear high notes and percussion effects. Each instrument should retain its true character and not interfere with the quality of other instruments.

Small combos do not require much dynamic range; if the recording is made at the proper level, the background will sound quiet, even on a machine with quite limited dynamic range. What really requires a good dynamic range is a recording of a symphony orchestra. If a machine handles the *crescendo* passages without distortion, and no background sounds are heard during *pianissimo* passages, then you really have dynamic range. This is a quite rigorous test, but one you'll need to make if you plan to play recordings of symphonic music.

SOME EXAMPLES OF CARTRIDGE AND CASSETTE SYSTEMS

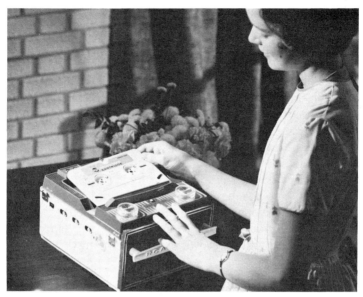

Early RCA recorder used large size cartridge shown being placed into operating position.

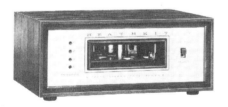

Heathkit Model GD-28 8-track stereo tape player uses standard NAB endless-loop cartridges.

Norelco "Carry-Corder 150" uses small reel-to-reel cartridge (cassette) developed by North American Philips. It is similar in principle to the RCA cartridge.

Wollensak M-4 stereo player/recorder uses small cartridges developed by CBS/3M. The stack shown provides 15 hours of continuous playing.

Lafayette Model RK-210 car stereo player/recorder which uses the Philips cassette.

Automatic Radio Model CFE-8001 car stereo player for playing 8-track NAB cartridges.

Allied Model 1125 stereo cassette player/recorder tape deck.

Ampex Micro 42 car stereo cassette player/recorder.

Wollensak 4800 Series stereo cassette player/recorder tape deck.

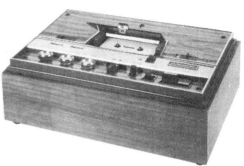

Roberts Model 808 stereo player/recorder system using 8-track NAB cartridges.

Craig Model 3108 car stereo player for playing 8-track NAB cartridges.

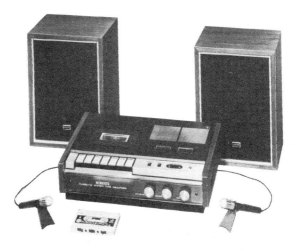

Roberts Model 100 stereo cassette player/recorder system.

Concertone Model 216-S stereo cassette player/recorder system.

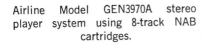

Airline Model GEN3970A stereo player system using 8-track NAB cartridges.

Norelco Model 2602 car stereo cassette player.

41

How about equalization? Every tape recorder must have a certain amount, but it will vary from machine to machine. If the specifications claim that the recorder has NAB or MRIA equalization, it probably does. The equalization of the recorder may not be too close to the standard—that is, although following the general pattern of response, it may not stay within the limits NAB and MRIA call for—but if the claim is made, chances are the performance is much better than in a machine where the equalization is not specified.

Machines that do not specify equalization give all kinds of quality. One will sound "boomy," or bass-heavy; another will sound shrill; another hollow. Almost every machine will give a totally different sound, even though the same type is played on all. Listen for the machine that produces a balanced (neither boomy nor shrill), smooth (neither hollow, harsh, nor squawky) sound.

If you want to make recordings of good quality, check the machine in this respect too. Several machines just above the lowest-priced bracket give quite a good performance in playing prerecorded tapes—but do a horrible job of recording a program. The trouble may only be maladjustment of the ultrasonic bias supply—or it could be poor circuit design. Either way, you'd do better to leave that model on the shelf and look for a better one.

To test the recording ability of a machine, first verify the quality of sound from a good tuner or from a phonograph record. Then try recording a program, at the proper record level, and see how it plays back. If the quality is below par, pass up that recorder. You may have to pay more than you expected, to get a recorder that can reproduce the original program with no distinguishable loss of quality. With patience you should find one at a more reasonable cost in which the difference is not too great.

COMPLETE RECORDER, OR DECK AND ELECTRONICS?

The electronics part of a tape recorder can be provided in various ways. Most manufacturers build the entire recorder as one unit, so you don't need to work out what's in it, or how to use it. Complete instructions, covering most of the things you can do with it, will be provided. A few manufacturers make only a tape deck, which is just the mechanism with its head mounts; the amplifiers (with an oscillator for the record function) must be bought separately. This is done so you can select the quality of amplifier that fits your needs (and pocketbook).

For what you are prepared to pay (except for the very cheapest, of course), a deck gives better quality than the composite recorder complete with electronics. This does not mean you are saving money by buying the deck and electronics separately. Although some high-fidelity preamps include an input for tape, the quality will not be as good as it will with a playback amplifier *designed for the heads on that deck*. And you will still need a record amplifier with its bias/erase oscillator. In all, you won't save money this way.

One advantage of the tape deck is its flexibility. It can be mounted wherever your fancy strikes you—as part of a wall assembly, upright, horizontal, or even at an angle. The electronics can be housed elsewhere if more convenient. You are not committed to finding room for the whole unit and providing yourself enough room to put on and take off tapes, make corrections, manipulate the controls, etc. You can even build-in the unit and have a really professional-looking installation.

Few people choose a car without giving at least some thought to the many ways in which cars differ. They decide what features are most important to them—appearance, comfort, handling ability, economy, performance, cost—and then buy one that fits their notion of requirements. Likewise, only a very few buy a Rolls-Royce. They cannot afford it; instead, they buy the car that best suits them for what they can afford. . . . It will repay you to follow the same advice in buying a tape recorder!

SUPPLY SOURCES

Most of the larger, professional-type machines, use the line voltage as the supply source. However, the growing

SPECIAL FEATURE RECORDERS

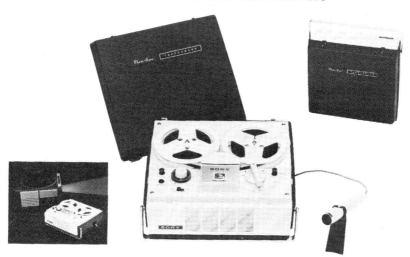

Sony/Superscope Model 211-TS photo/sync tapecorder.

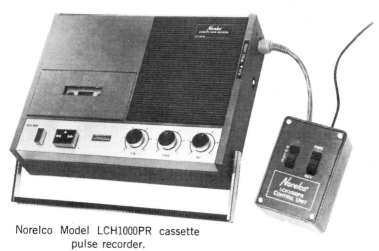

Norelco Model LCH1000PR cassette pulse recorder.

Lafayette Model RK-160 portable cassette recorder with a-m/fm radio.

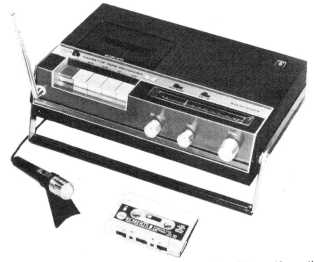

Roberts Model 525 portable cassette recorder with a-m/fm radio.

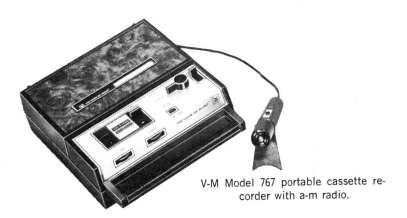

V-M Model 767 portable cassette recorder with a-m radio.

Norelco Model 2401A automatic stereo cassette changer, record/playback deck with Model CC6 continuous cassette circulator.

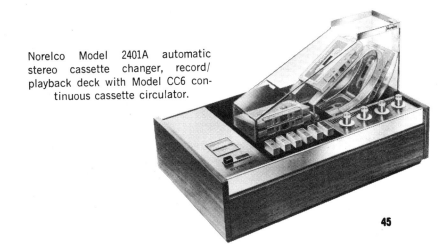

Ampex Model 1455 stereo recorder featuring automatic pickup/takeup and automatic reverse.

Ampex Micro 30 portable cassette recorder with a-m/fm radio.

Norelco Model RR-482 portable cassette recorder with a-m/fm radio.

Roberts Model 778X stereo combination reel-to-reel and 8-track cartridge player/recorder.

number of portables, and some of the smaller machines not intended specifically as portables, provide a choice of supply source. Some of them are battery operated. And some of them provide for alternative supply sources: battery or line voltage.

A few of the inexpensive imported models (not illustrated in this book) provide such alternative supply arrangements, but will not work on line voltage unless the batteries are inserted as well. This is because the line voltage provides a rectified supply that charges the batteries in series, while the machine uses different battery taps that are just not available if the batteries are not there.

The superior portables often provide batteries that are rechargeable over a long-term basis. This is intended to make the units suitable for field work, where they will generally be operated without the convenience of connection to a line socket. They can then be recharged overnight for use on the following day.

Making a unit rechargeable is not necessarily the same as providing it with dual source—line or batteries. If some units that are rechargeable are operated when the line connection is made for recharging, the result may be serious hum, because of the ripple current going into the batteries, and the fact that this provision is made for recharging, not as a mode of operation.

In choosing a tape recorder, you may want to know which kind of supply provision to get. This is why mention was made of the foregoing possible differences. Supply provisions are related to cost, especially in the less expensive recorders.

The least expensive recorder is one that uses batteries, especially when it is supplied without batteries. It does not have any internal supply provision. A recorder that uses line voltage supply must have transformer, rectifier, and filtering for all of the various supply voltages needed by the recorder.

The next simplest step is to use rechargeable batteries, which are fairly expensive, but when the manufacturer advertises "batteries extra" that fact does not show on the price of the unit. So watch for that, and check battery cost. Rechargeable batteries may cost from five to ten times the price of simple expendable batteries, as used in flashlights, which is why you do not see many rechargeable flashlights around!

A charging unit is much less costly than a supply unit with all the filtering necessary to run the machine without batteries. That explains the situation where a recorder with a battery charger produces a hum when the charger is connected while the recorder is operating. It is not simpler to "cure" than it is to buy a machine designed to work from either power source in the first place, which costs a little more.

When the earlier editions of this book were written, transistors had not been applied to tape recorders to any extent; a transistorized recorder was still something of a novelty. Now it would be difficult to find a new tape recorder that uses tubes—everybody has transistorized. This has made battery operation very much easier to achieve. With the tube-type amplifier, providing filament or heater current to activate the tubes took as much battery power as everything else put together.

Transistors avoid the need for filament or heater power; they also require much less operating power, the equivalent of B+ in tube systems. So a battery-operated recorder is far more feasible and economical than was ever possible in the days of tube amplifiers.

Every recorder must have some motors to drive the transport mechanism, and these require supply too. For the larger machines, some of them use ac motors, usually of one or other induction or synchronous type, while others use a rectified supply and operate on dc. Some of the more sophisticated machines use servo-controlled drive, and may use either ac or dc motors for this purpose. The choice will affect the electronics used to provide the servo control, a matter too advanced to discuss in this book. The relative merit depends more on the design than on a basic difference in type.

Earlier portables of the tube era provided spring-wound motors to save battery power, because the demand for those uses that could only be provided from batteries was so large. Now two factors have led to the universal adoption of battery-driven motors: The fact that other demands have been diminished, and motor design requiring much less battery power for this function has been improved in a lightweight, precision designed recorder. Also, greater battery capacity has helped.

Many people have asked me about this. Apparently, they have the impression that if they have a battery-operated recorder, a small change can make it line operated. Obviously, they base this assumption on the notion that the line socket is already available on the wall, whereas they have to buy a battery.

This false assumption ignores the fact that line voltage is ac, and has to be converted to the various smoothed dc voltages needed to operate the recorder before it can be used. Batteries provide the dc voltages naturally, with no conversion needed. So the true situation is almost the reverse of what many people assume.

An equipment designed for line-voltage operation can more readily be adapted to work from batteries than vice versa. The only possible disadvantage of this conversion is that an equipment designed for line voltage may not be designed with economy in mind, whereas an equipment designed for battery operation usually is. So a line-operated equipment, converted for battery operation, will probably use up batteries faster than one designed for battery operation. Or if rechargeable batteries are used, they will run down faster and need recharging more frequently.

However, for most readers, that possibility will be purely hypothetical—something they may talk about doing or having done, but when they find it is not really economic, they will abandon the idea.

Know the relative merits and have some idea of the relative costs. This, along with all the other differences between which you have to decide in choosing a recorder for your needs, may be used as a guide to decide whether you want one that is line operated or battery operated, one that can use both, or one that is rechargeable.

If you plan on extended use indoors, at a more or less fixed location, you should certainly get a line-operated model. For the relatively small extra, unless it is one of the more standard "professional" types, you may want to consider having one that can also use batteries, for those occasions when you may want to use it elsewhere.

If most of your work is on the move—reporting or some similar activity—a portable battery-operated model is a must. You may want to get more than one recorder, one for portable and one for indoor use. The portable saves weight by not having provision for line operation. And, if it comes with provision for battery recharging, this will usually be as a separate unit that does not have to be carried around.

3

Using a

tape recorder

It's most fortunate that tape can be used over and over again, instead of being "used up" every time something is recorded on it. Therefore, its costs you nothing to practice using the recorder to become familiar with its operation. If the trial recordings are no good or if you didn't want the program permanently recorded, the tape is not wasted —you can erase it and it will be as good as new.

In fact, when you first obtain a tape recorder, it is wise to be like a child with a new toy. In this way you can find out what it can do, and what mistakes to avoid later when you are taping more important material.

MICROPHONES

Monophonic recorders have always come with a microphone, although not one of very good quality—only good enough to "record to amaze," a trick outdated by the growing popularity of recorders. The earlier stereo recorders did not have microphones provided, mainly because mikes of good enough quality to do a creditable job for stereo were so expensive their inclusion would have priced the complete package out of the market. But the need for good mikes has given rise to the product, and now the mikes supplied with newer stereo recorders are good enough for quite a lot of effective stereo recording.

However, the critical stereo enthusiast may even want to get better microphones if he wants his recording to assume a quality more like professional recordings. Most of the supplied mikes are of the omnidirectional type. Even if their quality is quite good (smooth and level frequency response) they tend to pick up too much reverberation or echo effect, which can be avoided by the use of a more directional type of mike.

While microphones are still basically a product where you get what you pay for, the need for higher quality mikes at a price the home enthusiast can afford has produced a good crop of inexpensive mikes of acceptable quality. Even so, getting the mike you need is still more than "picking a good one." There are technical respects in which microphones differ, not connected with their quality, that can make a perfectly good mike unusable on your particular recorder.

A microphone has to be good, by which we usually mean it is distortion-free and has a smooth and level frequency response. Then also, it has to have the correct impedance and output level for the recorder's input. A wrong choice in these departments will not just "spoil" the recording (as would a poor mike); it will usually mean the mike is incapable of doing any recording at all. So, it is very important to see that you have these matters right.

Three kinds of mikes will be encountered with the type of recorder available for home use—dynamic, ceramic, and variable reluctance, or magnetic. Each type has different available properties, so it is important to know which you have and which kind your recorder can use. A fourth type, used professionally, is called a capacitor or condenser mike. This type is costly and its quality is not materially superior to good mikes of other types. Its use has other problems that make it inadvisable for the home user to give it a second thought.

Dynamic mikes include moving coils and ribbons. These mikes can have their coils wound, or they can be supplied with integral transformers that present any of the common impedance values: low, line (or middle), and high. Ceramic mikes are available only in high impedance. Lately some ceramics have been called "low impedance," but this is in comparison with other ceramics; they are still high impedance when compared with the other types. The ceramic high-impedance mike has different properties from other high-impedance mikes.

As a rule, the variable reluctance or magnetic mike is available only in high impedance. It does not have a very high quality, but is frequently the type supplied with inexpensive recorders because of its low cost and high efficiency. It is unlikely that you would want to buy this type, but you may find it is the type your recorder is already using.

With recorders that use tube amplifiers, low-impedance and line-impedance mikes have the disadvantage of requiring a well-shielded input transformer, and this is quite costly. Transistor amplifiers can inexpensively accommodate these impedances, so some of the recorders will be using them.

True professional equipment universally uses line impedance for microphone connections. To save cost, some equipment that is labeled "professional," with intent to indicate that the quality meets professional standards, may not use line impedance. If a transistorized version does use line impedance, it may avoid an input transformer. But, this means it is not of true professional standard because the transformer serves as precaution against hum and other spurious effects in big-system connections, which do not occur in a small-recorder setup.

Dynamic mikes are available with high-impedance connection and where this type is supplied with a home recorder, high impedance will be used unless the recorder is transistorized. Variable reluctance or magnetic mikes are also high-impedance mikes. The main disadvantage to high impedance is that too long connecting leads (over

SOME USEFUL MICROPHONES FOR STEREO RECORDING

Norelco Model EL 3752/01 contains two directional microphones in one case for stereo pickup.

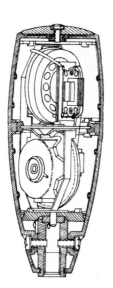

Internal view of the Norelco EL 3752/01 stereo microphone.

Reslo Mark III (two needed for stereo).

Shure Model 546 Unidyne III is typical of excellent range of dynamic cardioids.

Electro-Voice Model 666 dynamic cardioid.

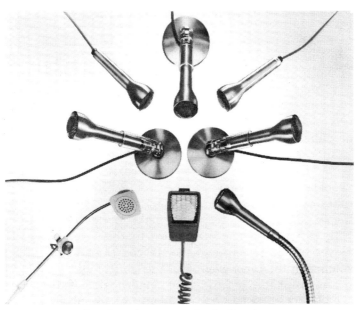

Low-impedance ceramics by Sonotone.

10 feet with standard cable, or 25 feet with special, low-capacitance cable) cause considerable high-frequency loss, making the quality "woolly."

Although ceramic mikes are also high impedance, this objection does not apply to them. In fact ceramic mikes are quite different in several respects: (1) their level is much higher than other mikes; (2) they must be connected to a high-impedance input designed for their particularly high impedance; (3) long leads affect them far less. Quality is not degraded, but some level may be lost if the leads are made very long (hundreds of feet). Ceramics do provide a low-cost answer to the problem, so many stereo recorders provide twin ceramics of quite acceptable quality.

If you find your recorder uses dynamic mikes of low, line, or high impedance, upgrading is relatively simple. Merely buy a better mike of the same impedance. You may buy a mike with a different directivity characteristic; for example, a ribbon mike which is bidirectional or a dynamic cardioid which is unidirectional.

If your recorder uses the variable reluctance type, upgrading is more difficult and quite often impossible, because the recorder's circuit is not designed well enough to allow improvement, even by an expert. In this case, you had better think in terms of a trade-in for a recorder of better quality.

If your recorder uses ceramic mikes, you may be able to replace them with better quality ceramics, or you can use another type by using a home-built transistor mike amplifier, which we describe in a later chapter.

Microphones also vary in sensitivity—usually, the better the quality, the lower the sensitivity. So you will probably need to compromise on one with moderate sensitivity consistent with a good, smooth response. To check the sensitivity and response, make a sample recording with the microphone about two to three feet from the source of sound, say a piano. If you cannot get enough volume on the tape, even with the control all the way up, the microphone sensitivity is inadequate. If the piano sounds tinny or otherwise unnatural on playback, the microphone does not have a satisfactory response.

For most recording jobs, a microphone with some type of directional pickup pattern is an asset. Two types of pickup patterns are cardioid and bidirectional. A microphone with a cardioid pattern (Fig. 3-1) is more able to pick up the sounds you want and to omit those you don't. But with a little extra care, a microphone with a bidirectional pattern (which is invariably a ribbon type) gives a better response (Fig. 3-2). The latter has two "live"

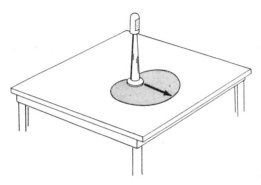

Fig. 3-1. The pickup patterns of a cardioid microphone. The arrow indicates the direction of maximum sensitivity; there is virtually no pickup from the back of the microphone.

areas instead of the one in the cardioid. Each area of the bidirectional pattern is a little smaller than that of the cardioid. But if you can get your sound sources properly located within these live areas, the quality is better than in a cardioid, and unwanted echo effects are not as prominent.

Another advantage of directional microphones is that you can control the pickup from sources of unequal intensity. For example, if a musical group includes loud instruments such as a trumpet, horn, or trombone, they will tend to drown out the others with normal "straight" miking. Your problem can be solved quite simply by using a microphone with a bidirectional pattern and placing the loud instruments to the less sensitive side of the mike. A few trial runs will show you where to place the microphone and instrumentalists for the best musical balance.

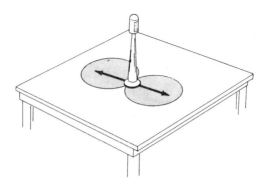

Fig. 3-2. The pickup pattern of a bidirectional (ribbon) microphone. There is approximately equal sensitivity in the direction of the two arrows, and practically no pickup at the sides.

The same is true of a conversational group. Where a number of people are talking, some invariably talk louder than others. You may not notice it too much until you play back the conversation; then it becomes quite conspicuous. Placing the microphone nearer the softer than the louder voices is the answer, of course. You'll have to be tactful; it seems ironic, but most soft-voiced people are microphone-shy, whereas most loud-voiced people insist on blasting right into it! If you can hide the microphone and casually maneuver your subjects into positions without being too obvious about it, you'll run less risk of offending the louder ones.

For stereophonic recording, the omnidirectional type usually supplied may pick up too much echo. The use of either a good ribbon properly placed to emphasize the "wanted" sounds and minimize the echo, or a cardioid (unidirectional) type, will help this considerably. Also, the smoother the microphone's response, the more effectively will it appear to offset unwanted echo effects.

RECORDING WITHOUT A MICROPHONE

For many kinds of recordings, you don't need a microphone. In fact, if you don't need one, it's best not to use a mike at all! You'll never get as good a recording, for example, by putting the microphone in front of your

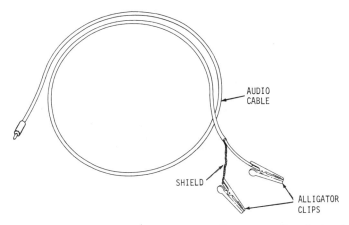

Fig. 3-3. A cable used to connect the radio output directly to the recorder input.

radio, as you will by connecting from the output circuit of the radio to the input jack on your recorder. To do this, attach alligator clips to one end of a lead. At the other end, attach a plug to fit the radio-TV-phono input socket of the recorder (Fig. 3-3). Hook the alligator clips to the speaker terminals. (Take care the clips do not touch one another and "short out" the signal.) If you get too much hum, reverse the alligator clips on the speaker terminals.

You can do the same with public-address systems, but you'll have to get permission from the PA system operator. Whenever you do this kind of recording, try to arrange for a "dry run" first, to make sure everything is operating properly and your record level is set correctly.

USING THE RIGHT LEVEL SETTING

The matter of having the right level is important with any recording, whether you use a direct connection or a microphone. If the level is too high, the recording will be distorted. If it is too low, the background noise will be excessive. Turning down the playback volume control will not remove the distortion; nor will turning it up remove the noise. Hence, you use the right setting on record, to minimize both distortion and noise at the same time.

SOME OF THE THINGS TO KEEP A RECORDER AND TAPE IN GOOD WORKING CONDITION

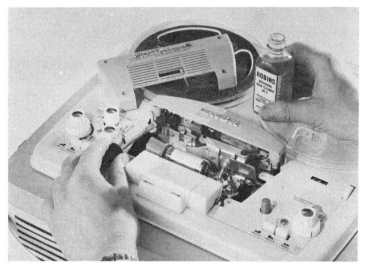

Using recording head cleaner.

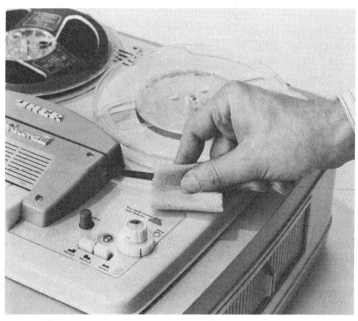

A tape jockey cloth cleans, lubricates, and puts a silicone film on the tape.

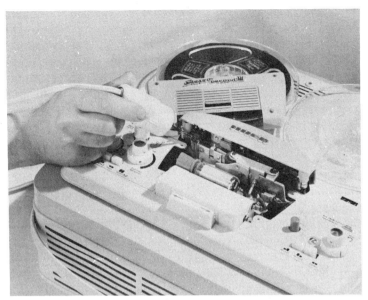

Using a tape head demagnetizer.

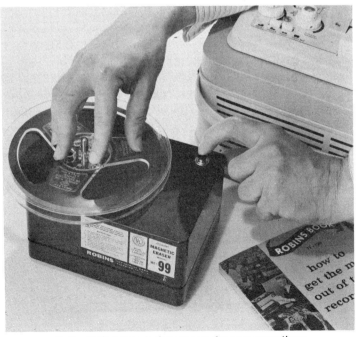

Bulk erasing tape before re-use for new recordings.

Most recorders have some kind of volume indicator. Some have a meter, the needle of which jumps back and forth as you are recording. Others have an indicator light (or lights) which flicker with various degrees of intensity. Whatever the machine has, don't take it for granted. The correct setting of the volume control varies according to the grade of magnetic tape you are using, and the recorder may have been set to indicate for a different grade.

When you make a "dry run" (as you should when taping any public function where something could go wrong), set the record level control so the meter needle reaches the zero mark on maximum upward swings only, or so the lights flicker in the manner prescribed in the operator's manual. Then play back your sample. If the recording is distorted in places, the level setting is too high and you'll need to find out how much lower it is necessary to set it to avoid distortion.

On the other hand, your first playback may be free from distortion but rather weak. If so, you might be able to use a higher level without running into distortion and, at the same time, get rid of any noise. Again, a few trial recordings may be needed before you will know how to interpret the meter reading or indicator light.

It is the peaks that distort, so pay more attention to how high the meter needle swings than to where its average position is (or how brightly the light flickers rather than how often, although the oftener it glows the higher the level, of course).

Having set the level during a dry run, don't assume it won't need changing. You may have set the level when there was little background noise—say, before the party "warmed up." The level will then need to be lowered. Or, if it's family speech-making you're after, Johnny may speak out quite boldly when you get him to try out, and then get a fit of nervousness and "swallow his tonsils" when it's time for the actual speech. When this happens, you'll need to bring the level up.

If you record weddings—incidentally, a tape of the ceremony makes a wonderful present for the newlyweds —you'll have to do some fast juggling of the volume control. The parson isn't nervous when he says his piece—he conducts weddings all the time—so you can leave the volume at a normal setting. But when the bride or groom timidly repeats the vows, he or she will barely be audible on tape unless you quickly turn the volume all the way up. But be sure to turn it back down before the parson says his next piece, or he'll blast the tape.

Another important factor for level setting is the kind of tape used. For a long while there was not very much difference in the level that different tapes would take. But now some superior tapes have extended the level that can be recorded. You want to use this extra level, when it's available, to get wider dynamic range (push the hiss level apparently lower).

TONE CONTROL

If the tape recorder has a tone control, you can sometimes use it to improve the sound. Be careful—if it works on both record and playback (sometimes the instructions omit to tell you this) and you adjust it for both record and playback, you may overdo it. On some machines, the control works only during record, which enables you to adjust for tonal quality while you record; the tone cannot be changed during playback. Others do the opposite; you can record without the tone control and then adjust what you hear when you play back. It is important to know which way the tone control works and then use it correctly.

You can readily tell (if the instructions don't) whether the tone control works on playback, by turning it back and forth to determine whether it affects the sound. To see if the tone control works on record, make a recording and turn the control during a certain passage, carefully noting at what point you do so. Then replay the piece and listen for any difference in sound during the passage.

MULTITRACK RECORDERS

There are many more things you can do if you have a multitrack recorder. For example, you can listen to one

SOME USEFUL ACCESSORIES
FOR RECORDER AND TAPE

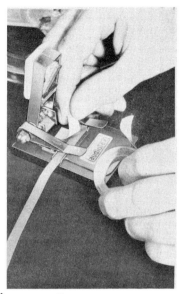

Using an Audiotex tape splicer.

Identifying tapes with Robins stick-on labels.

EDITall Type KS-3 editing and splicing kit.

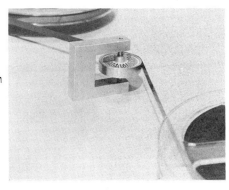

Checking for correct speed with a Scott strobe.

Robins reel-retaining holders for vertical operation.

Using a 3M tape clip to prevent spilling of tape when not in use.

track and, at the same time, record on another one. It would be a waste of time to give details here, because every recorder does this in a different way. As long as the recorder you have will do what you want, follow the manufacturer's instructions to make it do so.

You may have quite a bit of recording to do that is only single channel (mono) although you possess a stereo recorder. What do you do? Actually, you have a choice. You can ignore one channel and keep it blank for use on another occasion, which doubles the effective amount of tape on a reel, or you can, if the machine provides for it, tie the two channels together (monophonic) so that each channel records the same. At one time we would have stated that this will increase the dynamic range. It still will, of course, but the range is so good on many modern machines that any improvement is going to be difficult to detect.

CARE OF RECORDER AND TAPE

Having fun with your tape recorder means keeping it in tiptop condition. Like any other precision instrument, a tape recorder requires a certain amount of pampering. Dust it frequently, oil it regularly, and *never* let dust and oil accumulate on it. To a delicate moving part, dust is an abrasive. Also, should oil get on the tape, the quality of your recordings would be radically impaired.

Some recorders are easy to get into and clean; others require you to be a contortionist, unless you dismantle the recorder. If you do, don't touch anything that would interfere with the recorder's operation—tampering with a head, for example, could upset the delicate alignment of the head.

For various reasons, you may want to check the speed of your recorder—perhaps prerecorded tapes or your own recordings have too high or too low a pitch. The reason is that the speed is off, or for some reason it varies between record and playback. A big help in tracking down speed variations is a strobe that can be put against the tape, to show whether or not it is running at the correct speed.

Some recorders use a permanent-magnet erase head. By transfer via the tape to the other heads, they will also be permanently magnetized fairly quickly. To overcome this residual magnetism, frequent demagnetizing of the heads with a demagnetizer is necessary. A recorder with a well-designed bias/erase oscillator (which uses ultrasonic erase) will rarely need such attention if at all. However, it doesn't harm the heads to demagnetize them, and it's a wise precaution.

Sometimes the tape slides through the head path a little differently during record than it does during playback. This can cause poor alignment with the resultant deterioration in quality. If too severe, the high frequencies will disappear. Residual "edges" of sound track, left over from previous recordings that have not been erased entirely, can also result. A bulk eraser is the answer to this problem. But don't use one unless you want to erase everything on the tape. You cannot erase one track at a time with a bulk eraser—everything goes.

Splicing and editing are done for different reasons, but the method is the same in each. Splicing is done on tapes that have been broken. Editing is done when you want to rearrange recorded material into another sequence. The idea is to make the splice so smooth that it can't be heard. Needless to say, it's quite a chore without a tape splicer. If you have one (they don't cost much), you can make professional-looking joints with no difficulty at all.

With some stereo tape recorders, you can edit the tape without splicing it. These are the ones that permit you to play back, say, the left channel while recording on the right one. There is no reason why you cannot rerecord what was on the left channel and add something to it. Some recorders will let you do this only once; you cannot play back the right channel and record on the left one. With other, more versatile stereo tape recorders, you can record from either track onto the other. This means you can make another addition each time. Thus you can record a quartet—all four voices being yours!

4

Simple Things to do With

tape recorders

OFF-THE-AIR RECORDING

Recording programs from your own radio is perhaps the easiest thing you can do with a tape recorder. It also enables you to build up your own library of tapes. In the previous chapter, we told how to connect directly into the radio, instead of the less satisfactory pickup from a speaker. What we would like to emphasize here—if you don't want to be disappointed—is to make sure everything is operating properly before you start to record the desired program.

Obviously, you won't want to interrupt the program in order to go back and check the middle; you would lose part of the program. Once you start the recording, you cannot check it until the program is finished. Still, you don't want to continue, only to find out later you didn't get a solitary second of the program on tape. The only solution is to be ready ahead of time by taping some trial shots to be sure they're going on the tape and the level is set for minimum distortion and background noise. Then you can go ahead and tape your program, listening to it at the same time to know when it's finished. Of course, with a three-head machine you can monitor the tape all the time, instead of listening to the radio direct. But if you don't have a machine of this quality, you will have to check as thoroughly as you can before you start.

If you have serious intentions of building up a tape library of radio programs, here is another reason for buying a three-head machine. With a two-head machine, the quality you get will be inferior to that of commercial prerecorded tapes. On the other hand, the quality of recording you can make with a three-head machine will equal (or at least come close to) professional standards, and the playback quality will be better than that of a two-head machine.

However, if your interest in off-the-air recording is confined mainly to special events and you would rather buy prerecorded tapes for normally available program material, the less expensive, two-head machine may be quite satisfactory.

A GROWING FAMILY

The first sounds baby makes are a wonderful record to keep—if a bit difficult to get. To do so effectively, you will need to prepare for it. Baby may be making wonderful sounds—until you appear carrying a microphone and trailing a long wire behind you. Then baby promptly clams up, and you've lost your chance. With babies you have to be quick on the draw! Keep the microphone handy, connected, and ready to go.

You may think babies are ornery about having their voices recorded. Adults can be, too! They will chatter all day, saying things you'd love to have on tape. But the moment you walk in with a microphone, you might as well have pinned a "silence" notice on your back—everything goes quiet.

SOME WAYS OF USING A TAPE RECORDER

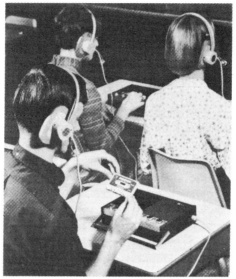

Courtesy North American Philips Corp.

Individualized study in the classroom permits students to progress at their own pace.

Courtesy 3M Co.

Recording junior's first words.

Courtesy Ampex Corp.

Transcribing notes onto tape while en route to or from an important business meeting.

Courtesy 3M Co.

Taped "living letters" reach across the miles.

71

Courtesy Ampex Corp.
Tape-recorder "secretary" records exact minutes of board or committee meetings.

Courtesy Ampex Corp.
Cassette car recorders have many uses such as taping on-the-spot impressions of special places and events during a vacation trip.

A good way to overcome mike fright is to get someone who is enthusiastic about hearing his own voice to say a few words while you record them. Play them back to him and as soon as the playback is finished, hit the record button again as unobtrusively as possible. In this way you'll capture his spontaneous comments. He'll undoubtedly make comments, such as "It sounds like your voice all right, but that's not *me!*"

Now you've got a starter piece. You can introduce it to the group by saying it's a piece of tape we took "just to hear how Harry's voice sounded." Before playing it, set the microphone up again, all ready to record. Don't say what you're doing. Your friends won't know that connecting the microphone isn't part of getting ready to *play* the tape. After playing Harry's voice, you again switch quickly to record and catch the spontaneous conversations of the others, too; then play them back. You'll undoubtedly be called a meany for pulling such a trick—but it works and they won't *really* mind.

Those are ways to overcome tape nervousness, but there are more serious uses for tape. In every family there are certain milestones—weddings, christenings, graduations, bar mitzvahs, and so on. A person looks back on events like these as turning points in his or her life. A tape of such memorable occasions makes a priceless memento. Because they can never be recaptured, deliberate preparation is a must.

If a public-address system is to be used for the occasion, it's a good idea to contact the public-address operator in advance. Don't just expect to walk in and make a recording; he'll be too busy to help you. Arrange to meet the public-address man before the event, so you can have your recorder set up, ready to go. Have a check-out beforehand, too, just to make sure everything is working all right. Events like this have no encore—you have to get it all the first time, or not at all.

OTHER OCCASIONS

Maybe a visiting notoriety speaks at your local social function or church, or a missionary tells of his absorbingly interesting experiences in the field. What a pity not to get them on tape. All that is really required is a little forethought. Suppose the event doesn't turn out to be a success? Well, the tape can always be re-used for something more interesting. But if the speaker has the audience on the edge of its seat and you didn't get his gems on tape, they're lost forever.

PORTABLE TAPE RECORDERS

The coverage of certain special events can be pre-planned, but for some occasions there isn't any opportunity to set up your equipment in advance. For example, you may be a member of the audience at a meeting or function, with no right to make special arrangements for your recorder. In this situation a portable tape recorder can be very helpful. Try to find a location in the room where you and your "one-eared friend" can hear the best.

There are other occasions which may be special but there's no way of setting up a standard recorder. A portable tape recorder can be very useful again. Suppose, for example, it's your young ones' first airplane flight.

You build up to it in advance: "Next week we're going to Elmsville by airplane." . . . "Tomorrow we're going on an airplane." . . . "Now we're going to the airport."

Interest is keyed up: "Ooh! Look, Daddy, is that our airplane?" . . . "Ooh, Daddy, there's another airplane." Johnny and Debbie are bubbling over with excitement as they breathlessly exchange comments. This special moment, captured on tape, will find a place along with photos and other mementos of a never-to-be-forgotten event.

The comments go on: "Daddy, we're off the ground! . . . *We're up in the air.*" . . . "Ooh, Daddy! We're going to have *dinner* on the plane." Other passengers are amused and there is a warm murmur of comments—maybe one from the stewardess—all grist for your taped "snapshots."

Preserve the spontaneity by carrying the recorder in your pocket or in a small case, and by concealing the mike on your person. If using a reel-to-reel type that uses small reels having limited tape capacity, be sure to carry enough reels with you in order to get the whole story.

Later, you can splice the shorter tapes onto a larger reel for playing on a full-size recorder. Also, remember to take some extra batteries.

If you use a reel-to-reel portable to record a special program, you may have difficulty in changing tapes without losing the continuity. The solution is to use two recorders, and to start the other just before the tape runs out on the first. This gives you time to reload while the second one is running. When you edit the tape, you can cut out the overlapping portions and finish up with a continuous tape of the event.

If you don't have two recorders, try to judge the best time to change tapes. When near the end of the tape, watch for a long pause in the program then change tapes.

If using a cassette-type portable, you will have more recording time available before you have to change tapes. Cassettes provide up to 60 minutes recording time on each side of the tape. Also, cassettes can be quickly and easily changed, thus helping you preserve continuity.

Portable tape recorders are not recommended for musical programs, unless it is the only way to tape something very special. Even then, you need to have the mike properly located, or else the quality will be poor. So the foregoing remarks really apply to speeches, discussion groups, or other such meetings. At many meetings you will notice that almost the whole front row is occupied by persons with portable recorders, who got there early to secure the most favorable position. (Unlike a photographer, who can always use a telephoto lens, you must get as close to the sound source as possible.)

Don't overlook the possibility of a portable tape recorder for "secret" use—either for fun, or for more serious purposes. Making a surprise recording of someone can be a lot of fun for everyone, if the person is "cutting up" and is completely unaware a recorder is nearby.

There's a legal angle to recording a telephone or face-to-face conversation. Without the other's *express* permission (i.e., written; or if verbal, in the presence of witnesses), you leave yourself open to a lawsuit charging invasion of privacy. Moreover, recorded telephone conversations must never be made in secret. In fact, a warning

SOME EXAMPLES OF PORTABLE RECORDERS

Allied Model 1150 cassette recorder.

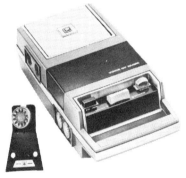

V-M Model 762 cassette recorder.

Lafayette Model RK-100 cassette recorder.

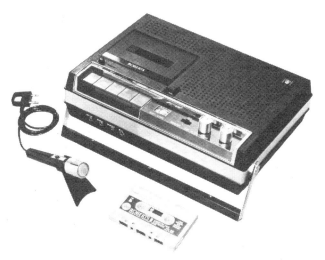

Roberts Model 80 cassette recorder.

Ross Model RE 3800 8-track cartridge portable stereo player.

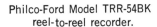

Philco-Ford Model TRR-54BK reel-to-reel recorder.

V-M Model 782 reel-to-reel recorder.

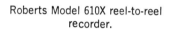

Roberts Model 610X reel-to-reel recorder.

Sony/Superscope Model TC-40 cassette recorder.

General Electric Model M8041A reel-to-reel recorder.

must be given at the start of the tape, in words to the effect that a recording is being made, and every 15 seconds an audible "beep" must be sent to remind the speaker the conversation is being recorded.

Tape recordings are often not acceptable as evidence in court. Not only are they sometimes classified as hearsay, but the courts are well aware of the ease with which tapes can be edited to distort the meanings of words by taking them out of context.

TAPE LETTERS

A rapidly growing use for tape recorders is sending vocal *letters*. The writer has been doing this for years, but more recently special mailers are available which make it easier. Some mailers hold a 3-inch spool of tape which will accommodate from five or six minutes to almost thirty minutes on each side. Mailers are also available for the Philips cassette tape cartridge. These provide up to sixty minutes on each side.

There is considerable flexibility in "writing" such letters. For example, you can fill only one side of the tape and leave the other side for a reply. Then, after you have "read" the reply, you can "read" your letter to refresh your memory. If you need the time, you can fill both sides.

Until you've tried it, you'll never realize the tremendous advantage of taped over written letters. The former are more intimate, more fun to receive than written letters, and—to the person for whom putting words on paper is a chore—more fun to send. Besides, a taped letter represents a familiar part of you—your voice.

For purely friendly correspondence or chitchat, a taped letter is a great timesaver, as well as being a more personal means of communication. Mailing a taped letter is not very expensive—it costs only a few pennies more than a written letter, and for the amount you can say, it is much cheaper than a written letter.

5

Practical Uses for

tape recorders

Before you had a telephone, you never knew what a wonderful servant it was; most people concede they would not know how to get along without a telephone. In some respects a tape recorder is similar—once you buy one, you'll never want to part with it. Instead of conveying sounds over distances, however, the tape recorder conveys sounds over another medium—*time*.

USING A TIME SWITCH

To tape programs off the air, you normally have to be there, so you can turn on the recorder before the program begins. But suppose you have to be out when the program comes on? You can still have the tape recorder make the recording. All you need is a relatively inexpensive time switch, to turn the radio and recorder on and off at the right time.

When using a time switch, allow enough time for the radio to warm up properly before the program comes on. Most radio receivers take about five minutes to warm up (although they may be "playing" in one or two).

To make the necessary adjustments, carefully tune to the desired station after the receiver has warmed up, and leave as is; also set the recording level so full output from the radio is a little below the level at which distortion begins. Allow a little leeway, in case the program level changes. It is better to have it go down, rather than up and cause distortion.

If the recording is to be fairly long, you may not want to start the recorder and turn on the radio at the same time, and thereby "waste" five minutes of tape. (The tape is not permanently wasted—it's just that you may not have space on the spool, in addition to the recorded material.) A way to overcome this is to use a thermal time switch that does not turn on the recorder until the radio has been on for a few minutes (Fig. 5-1).

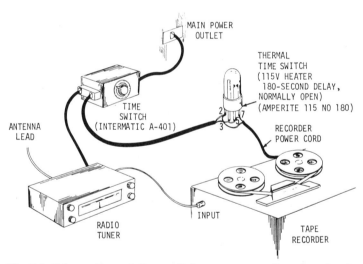

Fig. 5-1. Using a thermal time switch, so you can allow time for the radio to warm up before you start the recorder.

When you set something up like this to work by itself, check everything carefully, and have one or two dry runs. Make sure the time switch will come on exactly when you want it to. And if you use a thermal delay, make sure the time switch is on long enough to allow the delay switch to have the recorder running in time to catch the beginning of the program.

Also make sure the recorder is in the record position and its level is properly set. Then it will turn on the instant line voltage is applied by the delay switch, and will start recording as soon as the tubes warm up. (If the amplifier is transistorized, the warm-up is fast.)

If the recorder has an automatic shutoff, set the time switch to turn the radio off after the recorder has shut off.

COPYING TAPE

When you have a good tape, someone invariably wants a copy. This is fairly easy to do if he has a tape recorder too. Just place the two side by side and have yours play the tape while his records it. This is usually the best way, because the tape will be played on the machine that recorded it. If there are differences in quality between the two machines, you are used to yours and he to his. By letting each machine handle its own tape, these "normal" differences are preserved. If you change places, the differences are likely to be exaggerated.

There is another reason for doing this: The heads in both machines may not be perfectly aligned. By playing your tape on your machine and recording with his, the difference in alignment between the two machines will not affect the quality of the recording.

If both are high-quality machines of the three-head type, it may be worthwhile to disobey this rule. One machine may have the better playback head, capable of the highest-quality playback, whereas the other may have the better record head. In this event, the best tape will be made by using each machine for the job it does better.

SELF-CRITICISM

Many professional singers, comedians, public speakers, and others would not be without a tape recorder for quite a different reason. Any first-rate performer will be his own harshest critic—if he gets the chance to hear himself. This is not a matter of listening to yourself *as you perform*, because how you sound to yourself is different from how you sound to others.

To hear yourself as others do, you need a tape recorder. Entertainers do not want to waste valuable studio time polishing up their performances; they want to weed out any imperfections privately. So they have their own recorder at home to do this.

You may not be a professional, but only aspire to public speaking or other performing in an amateur way. In this event you have even more to benefit from in being able to hear your practice efforts. You'll be surprised at how much more spontaneous your planned "off-the-cuff" remarks will sound after you've had a chance to practice them and listen to the effect.

Don't try too long a speech all at one go; it can be too disheartening. If you record a long speech (say 15 minutes) the first time, you'll make so many *faux pas*, you may feel ashamed to present yourself in public. Take a little at a time; you'll gradually be able to expand the time you can put on tape at one stretch, until you begin to sound "pretty good."

PARTY GAMES

Some of the best fun to be had with a tape recorder is at parties. We've already suggested how to get over a person's self-consciousness about speaking into a microphone. Many of the older party games can be enlivened considerably with a tape recorder. For example, a form of charades can be played in which words are prerecorded and the players pantomime to them.

Or you can play one of those mystery games where individuals are invited into a darkened room, one at a time, without being given any clue as to what is expected of them. A concealed tape recorder (which is not difficult to do in a darkened room) picks up their facetious comments, to be replayed later.

In the replay, an emcee with a flair for this kind of thing can put an interesting "twist" to the comments. For example, preceding a remark of a young newlywed, the emcee may say: "Mrs. Newlywed, when her groom carried her over the threshold, commented . . ." and then the tape recorder comes in with Mrs. Newlywed's voice:

"Isn't it eerie in here." With some ingenuity, much fun can be had with this kind of game.

Of course, such recordings should be used exclusively for the party; to keep them afterward might lead to embarrassment. So erase the tapes as soon as the fun is over—and for your guests' peace of mind, let them know the tapes have been erased.

One more trick is good fun if you have a three-head tape recorder. Rig up a pair of headphones so the person speaking can hear from the playback head what he has just recorded into the microphone. Use the slowest speed the recorder provides.

This setup will make almost everyone stammer and try to race ahead of their speech in the most extraordinary manner. The effect of the headphones is to exclude the sound of their own voice as they normally hear it, and to bring it to them a fraction of a second later. Their first reaction is that they did not say a word or syllable; so they repeat it, only to find they've said the same thing twice. Then, to try to "beat" the double-image effect they talk faster than their normal pace and only aggravate their stammering. It's quite fun!

Onlookers who have not tried it are anxious to do so and are always confident they can do better—but they never can. So, by the time everyone has tried, there's been a lot of fun with it. One nice part about this fun is that the person who is trying enjoys the fun of his own efforts as much as he does listening to the others try.

MULTIPLE RECORDINGS

On the more serious side, you can use a tape recorder to produce your own multiple recordings. You can make like a one-man orchestra or choir (as Les Paul and Mary Ford did so wonderfully a number of years ago), or you can add simple echo effects.

To do the first, you need at least a two-track recorder. For adding simple echo effects while you are recording, one track will do if you have a three-head recorder. Plug the playback preamplifier output into the radio-TV-phono input, and plug the microphone into its input. The recorder must be the type in which both inputs (radio-TV-phono, and microphone) work at the same time but with separate volume controls, or the one in which the playback volume control can be used for controlling the echo separately from the record level control for the microphone.

The level fed back from the playback to the record head, to produce an echo effect, is very critical. If it's a whisker too high, the echo will be louder than the "original sound" and the recording on the tape will quickly build up to a deafening roar, without any semblance of a program. If it's not high enough, the echo will be imperceptible because the sound dies away too rapidly. So adjust the echo feedback very carefully, to get just the right effect.

Multiple recording is fun, but it can also be disappointing unless you have a really good recorder, or else restrict

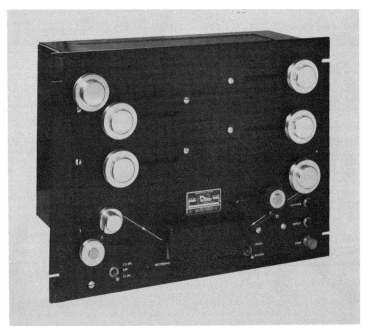

Courtesy Concertone, Div. of Monarch Electronics International, Inc.

A delay loop that can be used if you need more professional echo or reverberation effects than can be achieved with most stereo recorders.

85

yourself to a duet. The problem is that rerecording a program multiplies the quality defects of the machine (although the first rerecord may not be too bad).

On a three-head machine you have to rerecord the first track as you make the second one, to avoid the timing going seriously wrong. If you play back one track and record the other at another time, the two will be separated by a fraction of a second. This separation is due to the time the tape takes to travel from the record to the playback heads. When you play both tracks together, they won't be together—in musical time.

So the track you are playing back must be mixed with the new sound you are recording, after which the track you are playing back is of no more use because it is no longer "in sync." If you want to add another voice, all you have to do is play back this mixture to listen to it, mix it with the third (or fourth) voice, and record on the other track again.

If you plan to record harmony, we suggest you do the lower-pitched parts first because they suffer less from being rerecorded than the higher-pitched parts do.

Courtesy 3M Co.

Isolated booths equipped with earplugs and microphones enable students to hear the spoken language, record phrases in their own voices, then play them back for comparison with the original.

If you have a two-head stereo machine, you can avoid rerecording altogether, by recording the two effects separately on different tracks. This works only on a two-head machine because the playback head, by which you listen to the first track while you record the second, is exactly across from the record head on the second track. The two-head machine is the one more likely to have serious quality defects, so the advantage of not having to rerecord a number of times works out all right for making the best recording of a duet with yourself.

LEARNING A LANGUAGE

A final use for tape recorders, within the range of relatively easy things to do, is as an aid in learning a language. Get one of those language courses that have been recorded, and transfer the phrases to one track of your stereo tape. Leave enough spaces between phrases so you can repeat them later in your own voice. Now set the recorded track for playback, and the other track for record. In each space you can then record your own exercise. When you play the tape back, switch both tracks to playback and you'll be able to hear your own efforts right after the teacher's. In this way you can note your shortcomings. If you're not satisfied, switch the second track back to record and try again.

You could do this, with a monophonic recorder, by adeptly switching between record and playback. But the split-second switching makes it more difficult for you to concentrate on learning; also, you may erase some of the teacher's words as well as your own.

Besides being useful as a "teach yourself" aid, tape recorders are now being used extensively in schools for teaching languages and other subjects which must be practiced over and over. By this means each student can work on his own, in a soundproof cubicle. The teacher can then monitor each one's work in turn, and give the individual attention and advice so often missing in a lecture hall or large classroom.

6

Advanced Uses for

tape recorders

Earlier chapters have covered the relatively simple uses for a tape recorder. But magnetic tape is such a versatile recording medium that almost endless uses for it will come to mind. Some will be as uncomplicated as the ideas put forward in the other chapters. Others will need considerable ingenuity or extra equipment.

AUTOMATIC MESSAGE TAKER

A tape recorder can be rigged to take messages automatically when anyone rings your doorbell. There are a number of ways to do this. With a single machine you need a two- or four-track recorder with provision to have one track in the record and one in the playback mode. On the one set to the playback mode you can prerecord, at intervals, a message such as: "This is a recorded announcement. Mr. ——— is not at home at the moment; please leave your name and message. You have 15 seconds to do so, starting now." You can allow whatever time you deem necessary for the answer.

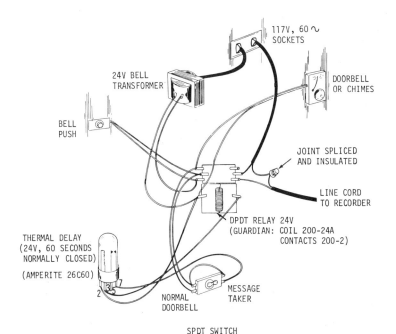

Fig. 6-1. Wiring diagram for system using single stereo tape recorder to take messages when you are out.

The tape recorder is set up so that when the doorbell is pressed, the playback unit says its piece over a speaker. Then a microphone picks up and records anything the visitor may say while the recorder is operating.

Fig. 6-1 shows a wiring diagram for a system using a single stereo tape recorder to take messages. The operation cycle is as follows:

1. Bell Push rings bell once and activates the relay and thermal delay switches.
2. Relay switches on the tape recorder, which makes the announcement and gives the caller time to reply on other track.
3. Thermal delay switch breaks circuit, letting relay drop out, switching off tape recorder, and disconnecting thermal heater. System resets for next caller.

If another caller comes before the thermal has reset from the previous caller, the doorbell will ring but the relay will not operate; when he tries again, after a short wait, the cycle will repeat.

The thermal delay allows 30 seconds for warm-up, 15 seconds for announcement, and 15 seconds for taking the message. A different thermal switch can be used if another timing is needed.

If you use a transistorized tape recorder, the doorbell and relay can switch the recorder on directly, and a thermal delay switch and the same relay will switch it off after the announcement has been played and the allotted time for recording has elapsed. To prepare the machine for taking messages, you record a number of announcements on the track by using the same time sequence for each. Allow a little time margin at both ends, in case the thermal time switch does not repeat its timing exactly.

If you use a recorder with tube amplifiers, it will be well to save tape by using a second delay switch to turn on the recorder motors. The doorbell will switch on the amplifiers immediately; then the time delay will allow the amplifiers to warm up before starting the motors. Finally, a second delay switch, connected in the motor circuit, will shut everything off.

A wiring diagram for a system, to avoid the loss of tape space during warm-up, is shown in Fig. 6-2. The operation cycle is as follows:

1. Bell Push rings bell once and activates the relay and thermal delay No. 1.
2. Relay switches on tape recorder, but not its motor.
3. Thermal delay No. 1 starts the motors and activates thermal delay No. 2; this makes the announcement and gives caller time to reply.
4. Thermal delay No. 2 breaks circuit, letting the relay drop out and switching everything off.
5. The thermals cool off, resetting for next caller.

The 30 seconds of thermal delay No. 1 allows time for the amplifiers to warm up; the 30 seconds of thermal delay No. 2 allows time for the announcement and for taking the message. Different thermals can be used if either of these is insufficient. In this circuit and Fig. 6-1,

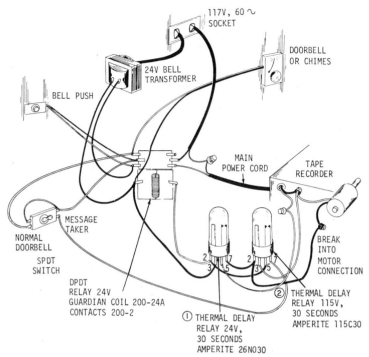

Fig. 6-2. Wiring diagram for a system to avoid the loss of tape space during warm-up period.

the spdt switch enables you to deactivate the system and operate the doorbell normally while you are at home.

With this arrangement, there is always the problem of being sure everything is synchronized. Allowing a margin may not be enough if you have many callers. If you allow five seconds for error, this may be enough for two or three calls. But with twenty calls, the announcement could come through just before the machine shuts off again, or the machine could start up in the middle of an announcement.

To make a more reliable system, you need two recorders—one to make the announcement and one to take the messages. On one you have an endless loop with the announcement recorded just once. A small piece of foil, attached to the back of the tape, serves as a contact to simultaneously stop the announcement recorder and start the message recorder.

A wiring diagram of this system is shown in Fig. 6-3. The operation cycle is as follows:

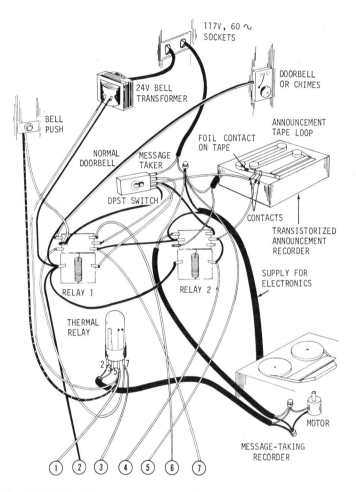

Fig. 6-3. Wiring diagram for system using two recorders—one to make the announcement and one to take the messages.

1. Bell Push rings bell and activates relay No. 1.
2. Relay No. 1 switches on transistorized announcement recorder containing tape loop that will repeat the same announcement any number of times; the circuitry of the message-taking recorder is also switched on, but not its motors.

3. Contact foil on announcement tape activates relay No. 2.
4. Relay No. 2 switches off announcement recorder and switches on the motors of the message-taking recorder, at the same time activating the thermal relay.
5. After a prescribed time, the thermal relay breaks the circuit, deactivating relays No. 1 and No. 2.
6. When the thermal relay has cooled off, the system will reset for taking the next message.

Relays No. 1 and No. 2 use Guardian coils 200-24A and contact assemblies 200-2; the thermal relay is *Amperite* 115C15 (for 15-second messages; others use corresponding delay times). The dpst switch deactivates the system and allows the doorbell to function normally. Terminals 1 through 7 allow time markers to be inserted with the addition of the parts in Fig. 6-4.

A transistorized recorder will work the instant it is switched on. You can arrange the loop length to allow the amplifier in the main recorder sufficient time to warm up before its motors start. A thermal delay switch, which comes into the circuit at the same time the motors of the message recorder are switched on, turns everything off

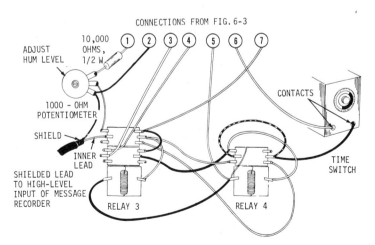

Fig. 6-4. Wiring diagram of additional parts for adding time indications at desired intervals.

(by breaking the main relay circuit, which closed when the doorbell rang), thus setting the system so it is ready to take the next message.

The system just described requires more equipment and is more complicated to set up; but it's easier to make ready, and will take a number of messages, dependent on the capacity of the tape in the message recorder. If you set it to give 15 seconds per message and have an hour's playing time on the reel of tape, 240 messages can be recorded before you need to reload. That should be enough even for a vacation!

Don't try to hook up a recorder to your telephone—the company does not allow any such devices to be attached to their equipment. Actually, the telephone company will rent you a recorder for a modest monthly fee (about the same cost as an answering service).

TIME INDICATION

If you want to provide indications of the times at which calls were made, a time switch and extra components can be added to the hookup in Fig. 6-3. These can run the tape at specified times, set up in advance on the time switch, and switch over from microphone to a controlled hum input. The wiring diagram of the additional parts needed for adding time indication is shown in Fig. 6-4. The operation cycle is as follows:

1. When the time-switch contacts close (no message being taken), relays No. 3 and No. 4 are activated, also activating relays No. 1 and No. 2.
2. Relay No. 3 switches a hum input to the high-level input of the message taker; relay No. 2 prevents the announcement from being given, and also starts the message-taker motors immediately.
3. After the prescribed time for the thermal relay to operate, all relays except No. 4 drop out and the thermal cools off, ready to take another message. Relay No. 4 remains activated, preventing a second hum time marker from being given before the time-switch contacts open.

4. When the time-switch contacts open, relay No. 4 drops out, resetting the circuit to receive the next time marker.

If the system is taking a message when the time-switch contacts close, relays No. 3 and No. 4 will not be activated until the completion of the message.

Relays No. 3 and No. 4 use Guardian coils 200-115A; relay No. 3 uses contact assembly 200-5; relay No. 4 uses 200-2; the time switch can be an *Intermatic* A221 for daily time markers, or a T960 for markers used at more frequent intervals.

The advantage of recording hum as a timing indicator on tape is that it stands out as a tone when you run the tape at fast forward through the recorder. Thus, if you record messages for each day of the week, you can run the tape and count the beeps, one for each day, to find the messages received on any one day. Or you can have the time switch make the beeps at more frequent intervals. But by using hum, the tone comes out as a deep, 60-Hz hum at normal playing speed, or a strong high-toned beep at fast forward or rewind. Finding the beeps among the messages is very easy.

The author sometimes uses a tape recorder for dictating articles and books. The dictated material may be stored on the tape for days, until it can be transcribed. If a number of items are dictated on the same tape, it helps to have a quick way of finding each one. If you are making permanent tapes with multiple programs, it is easy to splice in colored leader tape so the program separations can be easily seen. But you don't want to cut tape that is intended for repeated use, as in dictation. So the author has found the audible hum-marker method very useful.

To record the hum marker on the tape, just switch to radio-TV-phono input and stick a small screwdriver into the jack while touching the metal shaft of the screwdriver with your finger. Then, when you're searching for the marker, just use fast forward or rewind and the hum will be heard as a beep when it goes by. Thus, you don't have to keep stopping to see whether you've got to the place, or passed it. For a visual indication, adhesive tabs can be used (Fig. 6-5).

Courtesy GC Electronics

Fig. 6-5. Using adhesive tabs to mark the tape.

HOME-MOVIE SOUND

Sound can be added to your home movies by using a tape recorder. The possibility of doing this captured home-movie fans almost as soon as tape recorders made the scene. At first, for all except the "do-it-yourselfer" who would undertake a very ambitious project indeed, the extent of "sound" provided was merely appropriate background music: Synchronized sound and picture were not easily possible.

Some early experimenters were successful with a variety of systems, but they were too complex for any but the most sophisticated experimenter. Then came some systems that were presented in the earlier editions of this book, which are repeated here mainly to show how it *was* done. In those days, it seemed remarkable, because anything better was beyond the range of the home-movie maker. Now, in retrospect, it looks crude.

Courtesy Paillard, Inc.

Fig. 6-6. The arrangement for using a synchronizer between a projector and any tape recorder for synchronized sound movies.

The first method enabled any tape recorder to provide the "sound track" by using a synchronizing device to hold the projector speed exactly in time with the tape recorder.

One such device was the Paillard-Bolex *Synchromat* shown in Fig. 6-6. It was quite a simple mechanism. After the tape passed the heads, it went through the synchronizer before being wound onto the take-up reel. In doing so, it passed over a pulley and pinch roller like that used on the tape recorder to maintain a constant speed, but this one was driven by a flexible drive from the projector. Between this drive and the one for the tape recorder, the tape passed over a movable pulley.

As long as the recorder and projector were running at the same speed, the pulley kept the tape taut. If the projector started to slow down and caused the picture to run behind the sound, the tape began to slacken and the pulley moved back. In doing so, it speeded up the projector and tautened the tape again.

If the projector ran too fast, the pulley moved forward, cutting down the projector speed and again bringing the tape back to its proper length of loop. When the tape recorder stopped, the tape pulled the pulley to the extreme end of its travel and switched the projector off. When the tape recorder started, the pulley was allowed to move back, starting the projector.

In this way, exact synchronism was maintained, including stops and starts. All that was needed, to be sure the synchronism was right, was to have marks on both film and tape, so they could be started "in sync" at the beginning of each picture.

Making the sound track was just the same as playing it, except that you operated the recorder in record instead of playback. It was best to make the sound for fairly short pieces at a time, rehearsing each piece a few times before you recorded it. However, if you weren't satisfied with the final "take," you could always do it over.

The newer method presents no synchronizing problem. In most large cities there are laboratories which will stripe movie film with a magnetic track after you've had it developed. You can even have your old movies striped in this way. Your movie film shop will usually tell you the name of such a place, or have it done for you.

From there on, the procedure was at first quite similar to the one just discussed except for not having to worry about the synchronizing. The Paillard-Bolex *Sonorizer* tape recorder (Fig. 6-7) provided a method of making such a recording. It could be fitted to practically any projector that maintained a reasonably constant speed. (A projector could fluctuate appreciably in speed and not affect the picture, yet sound "warbling" would be objectionable). With the *Sonorizer* or any system using sound on film, the projector motor controls the speed, not a separate tape recorder as in the other method.

Courtesy Paillard, Inc.

Fig. 6-7. An add-on feature enabling a "silent" projector to be used with magnetic sound-on-film movies, both for making and playing the sound track.

As well as having the heads specially mounted so they can be placed in the film path between feed spool and intake sprocket, the electronics in the *Sonorizer* were somewhat different from most tape recorders, to make the job easier.

In addition to the usual record and playback positions, the function switch has a "superimpose" position. In record, (as with any other recorder), whatever is already on the tape or magnetic film track is erased; but in the superimpose position, the tape is only partially erased and the new sound could be recorded "on top" of it. This happened only while the button on the microphone was pressed; at other times the superimpose position functioned exactly like the playback.

This enabled the sound track to be made in successive "runs." First, background music could be recorded, then perhaps, sound effects. The sound effects would only override the background music, not erase it. Finally dialog, which will override the previously recorded material where voices have to cut in, was recorded. If desired, the final addition could be commentary. Every time the commentator pressed the mike button, the other sound faded down and his voice would cut in.

Adding background music was no problem; you just picked the passages you needed, making sure they were long enough, and dubbed them in. Sound effects, such as waves at the seashore, do not require precise synchronization (as long as the sound appears at the right place during the picture), because there are many sound waves striking your eardrums at once, and the sound you hear may not be coming from the wave you are looking at. You could dub-in these sounds from sound effects records, which have become popular since the advent of high fidelity.

Dialog needs quite careful synchronization, no matter which recording system you use. All the mechanical part of the synchronizer can do is make the synchronization as good as you record it. If you record it wrong, the synchronizer will make sure it is always played wrong too.

This is why it is best when using that method to record in fairly short "takes," and to rehearse each one a few times. The persons who are talking must watch the picture, to find out the exact *timing* of *what* they are to say (by lip reading). After a few dry runs, they will be able to synchronize their voices with the lip movements on the screen. Then they are ready to permanently record that piece "in sync."

Don't underestimate the importance of synchronizing dialog. If only one syllable is half a second too early or too late, the fact will be quickly apparent. Something will seem unnatural, like a ventriloquist whose voice is not in time with the dummy's movements. When synchronization is perfect, an illusion that the sound really comes from the person on the screen is achieved. But any momentary break in synchronism will destroy that illusion.

Not too many people use that method any more since the camera people introduced systems to enable anyone to make their own home movies with "live" sound. The picture is made by the camera, and the sound is recorded on a recorder and synchronized during the taking process. The system may merely have a microphone for the cameraman to make his own commentary, and pick up sound associated with the picture almost as "background," or it may have a more elaborate provision for extension microphones, so more professional sound movies can be made.

In any event, the sound track that needs synchronizing with the picture is made at the same time the camera "rolls." Then a processing agency, after the picture has been developed, transfers the sound from the recording to a stripe put on the film after the picture is finished. Then the final, finished striped film is presented on a sound projector that comes as an integral package, thus avoiding the cumbersome threading shown in Fig. 6-7.

If you want to add background music or other sound not obtained from the original "take," this can be added by arrangement with the processing agency. As this kind of film is a little more costly than that used for the silent-type movie, it is easier to make arrangements that depart somewhat from the mass-processing used for picture only.

Slide Presentation Commentaries

A more "natural" use for a regular tape recorder is to provide a recorded commentary for a slide presentation. Many modern slide projectors incorporate an electrical remote-control system for changing from one slide to the next. With such a projector, it is only a short step to make the whole operation automatic. All that is necessary is to have the magnetic tape carrying the commentary provide the electrical impulses planned by the lecturer. Then the whole presentation can be given, without requiring the presence of the lecturer. The necessary electrical impulses can be derived in a variety of ways.

One of the earliest setups (Fig. 6-8) used a mark on the back of the tape. It used an add-on unit to detect the mark that was made with a conducting pencil. The methods in use today utilize various signals recorded on the tape, either on a separate track, or suitably separated by audio discrimination on the same track.

Courtesy General Techniques, Inc.

Fig. 6-8. Setup for using the Mark-Q-Matic slide synchronizing device. In foreground is the special pencil for marking the tape, and the slide projector is to the left of the recorder.

A problem with using the same track, which carries voice or other program, is that it is difficult to prevent voice tones from activating the slide movement when it is not wanted. A form of signal, which will not appear in normal voice or music (if background music is used), must be devised. Preferably, it should be inaudible, or at least unobjectionable, which poses limitations.

Some sophisticated systems, using the same channel for signals like this, have been devised. However, as stereo recorders are now commonplace, a simpler method is to put the program sound—voice and music—on one track, and the slide-changing signals on the other.

TRANSISTOR MICROPHONE AMPLIFIER

If you want to use a better microphone that does not have enough gain, it is fairly easy to construct your own microphone preamplifier. Fig. 6-9 shows the construction details. This preamp provides an input suitable for a line-impedance microphone and it has an output level that will be adequate for a recorder intended for a high-level ceramic (or even magnetic) mike. It provides more than enough gain for most purposes, and the recorder's gain control can be used for control.

The resistors are standard ¼-watt resistors, the transistors used are easily available 2N109s, and the coupling capacitors are minilytics, only 9/16-inch long by ¼ inch in diameter. The whole assembly, when wired to a phone socket and plug and made to the dimensions shown, can

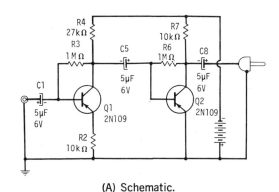

(A) Schematic.

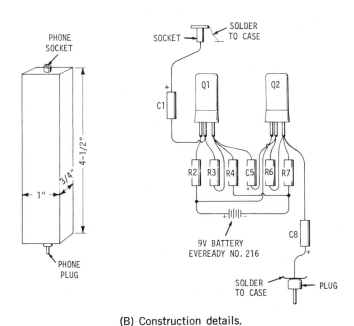

(B) Construction details.

Fig. 6-9. Microphone preamplifier.

Courtesy 3M Co.
Radio stations record programs on tape for broadcasting at a later time. Here, a college choir is being taped.

Courtesy 3M Co.
A special type of recorder used in satellites to record man's first words from outer space.

be enclosed in a metal case cut from scrap tin cans. The inside of the case should be lined with adhesive insulating tape before assembly (and after soldering together).

If desired, a switch can be wired into the battery circuit. An alternative is to make a panel of the case removable (about 2 inches high on one of the 1-inch sides) for inserting and removing the battery. Thus, the battery can then be put in just before recording is started, and removed when the job is done. The drain from the battery is less than half a milliamp, so it will not "flatten" quickly if left on for long periods by mistake. Because it is battery operated and totally enclosed in shielding by the tin can, the amplifier is very quiet in operation and quite hum-free.

This book has covered quite a collection of things you can do with a tape recorder. It could not possibly tell everything that will ever be done with tape recorders, because their uses are as endless as man's collective ingenuity to find more. We have concentrated on those the average person can do, or is likely to want to do.

We haven't even mentioned some of its most important uses: for computer memories, storing television programs for transmission later, or for sending overseas for transmission elsewhere, not to mention uses associated with data transmission to and from man-made satellites. If you're seriously interested in these other applications of tape, we don't recommend your trying them on your own. The better way is to get a job with a company specializing in that kind of work. There is always room for more people in such a rapidly expanding field.

The man in the street, home, or office is going to be seeing more and more of tape as time goes by. Recently, we learned of a company that is putting out a series of small cassettes that can act as your personal tour guide for Europe. If you set the recorder going and travel at the designated speed, it will tell you what you are passing and where to look. If you stop, you merely stop the recorder.

That will certainly be an inexpensive tour guide. The only thing is, you can't ask it questions. But then we have encountered human-type tour guide situations where asking a question was equivalent to replaying the last few inches of tape. So what's the difference?

We have all seen movies and TV shows where fake recorders perform every imaginable task in a futuristic world. Maybe that world is not so far away as we think.

It would be difficult to state what modern-day development has had the most widespread effect, but it is certain that magnetic tape, and the machines that use it, come high on the list.

Glossary

A

Alignment. Adjustment of the heads so they are correctly positioned to record or play the right track and at the correct angle (see Azimuth).

Audio. Electrical voltages and currents corresponding to sound waves; also the magnetic fluctuations on the tape corresponding to sound waves.

Azimuth. Having the *head* and tape at the correct angle to one another so the line formed by the head *gap* is exactly perpendicular to the *track* along the tape.

B

Bias Oscillator. The electrical circuit that generates the *ultrasonic bias* current; usually the same circuit provides *erase* current.

C

Capstan. The precision cylindrical drum or spindle the tape is pressed against, the purpose of which is to keep the tape moving at a constant speed.

Cartridge. A package containing tape, usually in an endless spool, designed to be inserted in a slot in the tape recorder that complements it. This tape plays all tracks in the same direction.

Cassette. A package containing tape, usually on two spools, designed to be inserted in the tape recorder without removal from the package. The tape has two "sides," an equal number of tracks being played from each end.

Channel. A complete program of sound, carried by an amplifier or other electrical apparatus, which will be recorded by a single set of magnetic fluctuations on one sound *track*. Stereo recording invariably needs two (or more) channels.

D

Deck. (1) An assembly containing the *transport* mechanism and *heads*, but not the *electronics* which are contained in a separate package; (2) the same part of a composite recorder.

Dub. To copy program from tape or a phonograph record onto a fresh piece of tape.

E

Editing. Rearranging program material on tape. This may be achieved either by cutting and *splicing* the tape in a different sequence, or by dubbing it, piece by piece, on a fresh tape.

Electronics. The part of a tape recorder that contains the amplifiers, equalization, and bias oscillator.

Equalization. A deliberate modification of the amplification accorded to different frequencies, so the overall reproduction of sound maintains its correct musical balance, resulting in minimum loss of quality.

Erase. The process of removing the magnetic recording from a tape before a new recording is impressed on it.

F

Flutter. A distortion occurring in a recording due to nonuniform speed of the recorder.

G

Gap. The minute magnetic slot in the *head* which comes in contact with the tape, and is responsible for impressing program on the tape, or retrieving it during *playback*.

H

Heads. The assemblies that transfer the electrical currents onto the tape as magnetic fluctuations during *record*, and retrieve them as electrical voltages during *playback*.

I

Interlock. A device designed to make it impossible to accidently record or erase a tape that is intended to be played. Also a device to prevent breakage of the tape by too sudden a stress.

L

Level Meter. A meter that indicates the electrical intensity of the *audio* being recorded or played back.

M

Monitoring. Playing back the recording immediately after it has been recorded, to see if the quality is satisfactory.

P

Pause. A control that allows the recording process to be interrupted momentarily, while a button is held depressed, without involving the use of an *interlock*.

Playback. Retrieving the program recorded on the tape, so it can be heard.

Prerecorded Tapes. Tapes bought with program already recorded on them, ready for playing.

R

Record. The process of impressing program on tape, in the form of tiny magnets that can later be used to reproduce the original sound on *playback*.

S

Safety Button. See Interlock.

Splicing. Joining tape together (either when *editing* or repairing broken tape) so it will play without a break.

Superimpose. To record a program over the top of another, without erasing the first program, so both are on the tape together.

Synchronization. The timing of two parts of a program, so they are identical in time to within a small fraction of a second. The parts being synchronized may be different sounds being synthesized into a single program, or sound and its associated picture.

T

Track. The width of magnetic oxide on the tape, occupied by magnetization corresponding to one channel of recorded sound. *Two- or four-track* refers to the number of such tracks occupying the total width of a tape.

Transport. A collective name for all the parts of a tape recorder that are responsible for controlling the movement and handling of the tape.

U

Ultrasonic Bias. An extremely high-frequency signal (above the audible range) that is combined with the audio signal to be recorded, for the purpose of improving recorded quality.

W

Wow. A term that denotes a change in pitch of the reproduced sound, resulting from a periodic, fairly slow fluctuation in speed during either record or playback.

Index

A

Accessories for recorder and tape, 64-65
  editing and splicing kit, 65
  identification labels, 64
  reel-retaining holders, 65
  splicer, 64
  strobe, 65
  tape clip, 65
Amplifier, transistor microphone, 102-105
Audible time indication, 94-96
Automatic message taker, 88-94

B

Bias, high-frequency, 15-17
Bidirectional microphones, 56-58
Brakes, 11
Bulk eraser, 61

C

Capstan, 8
Cardioid microphones, 56-58
Care of recorder and tape, 66-67
Cartridge systems, 9-10
  CBS/3M type, 9-10
  examples of, 38-41
  NAB type, 9-10
  Philips cassette type, 9-10
  RCA type, 9-10
Ceramic microphones, 52-56
Cleaner, tape head, 60
Clip, tape, 65
Cloth, jockey, 60
Complete recorder, or deck and electronics?, 42-43
Commentaries, slide presentation, 101-102
Controls, 24-25
Copying tape, 82
Cross-field heads, 17

D

Demagnetizer, tape head, 61
Dynamic microphones, 52-56

E

Echo effects, 84-85

Editing and splicing, 67
  kit, 65
Electronics, 21-24
Equalization, 23
Erase
  bulk, 61
  heads, 14-15
  high-frequency, 14-15
  permanent-magnet, 14
Examples of cartridge and cassette systems, 38-41
Examples of professional-type equipment, 34-35
Examples of special feature recorders, 44-46
Examples of ways of using a tape recorder, 70-72

F

Flutter, 8

G

Growing family, using a recorder with, 69, 73

H

Heads, 11-21
  cleaner, 60
  cross-field, 17
  demagnetizer, 61
  erase, 14-15
  playback, 18-19
  record, 15-17
  tracks and alignment, 19-21
High-frequency bias, 15-17
High-frequency erase, 14-15
High-impedance microphones, 53, 56
Home-movie sound, 96-101
How many heads ... motors?, 32-33

I

Identification labels, tape, 64
Indicator, volume, 62

J

Jockey cloth, tape, 60
Judging quality, 37, 42

K

Kit, editing and splicing, 65

L

Language learning with a recorder, 87
Letters, taped, 79
Level setting, using correct, 59, 62-63
Low-impedance microphones, 53, 56

M

Magnetic Recording Industry Association (MRIA), 23
Magnetic sound-on-film movies, 98-101
Message taker, automatic, 88-94
Microphone preamplifier, transistor, 102-105
Microphones, 51-58
    bidirectional, 56-58
    cardioid, 56-58
    ceramic, 52-56
    dynamic, 52-56
    high-impedance, 53, 56
    low-impedance, 53, 56
    magnetic (variable reluctance), 52-56
    stereo, examples of, 54-55
    unidirectional, 56-58
Movie sound, home, 96-101
Multiple recordings, 84-87
Multitrack recorders, 63, 66

N

National Association of Broadcasters (NAB), 10, 23

O

Off-the-air recording, 68-69

P

Party games, using a recorder for, 83-84
Permanent-magnet erase, 14
Playback heads, 18-19
Portable tape recorders, 74-79
    examples of, 76-78
Preamplifier, transistor microphone, 102-105
Price consideration, 26-27

Professional-type equipment, examples of, 34-35
Push buttons or levers?, 33, 36

Q

Quality, judging, 37, 42

R

Recorders
    care of, 66-67
    heads, 15-17
    portable, 74-79
        examples of, 76-78
    professional-type, examples of, 34-35
    special feature, examples of, 44-46
    typical stereo, 30-31
Recording
    multiple, 84-87
    off-the-air, 68-69
    without microphone, 58-59
Reel-retaining holders, 65
Reel to reel, cartridge, or cassette?, 27-29, 32
Reverberation effects. *See* Echo effects

S

Self-criticism, using a recorder for, 82-83
Slide presentation commentaries, 101-102
Sound, home-movie, 96-101
Sound-on-film movies, 98-101
Special feature recorders, examples of, 44-46
Speed, tape, 10
Splicer, tape, 64-65
Splicing and editing, 67
    kit, 65
Stereo
    microphones, 54-55
    recorders, typical, 30-31
Strobe, tape, 65
Supply sources, 43, 47-50
Synchronized sound movies, 96-101

T

Tape
    bulk eraser, 61
    clip, 65
    copying, 82
    editing and splicing, 67

Tape (Con't.)
  kit, 65
  head cleaner, 60
  head demagnetizer, 61
  identification labels, 64
  jockey cloth, 60
  letters, 79
  speeds, 10
  splicer, 64
  strobe, 65
  tracks and alignment, 19-21
Tape recorders
  considerations in choosing, 26-50
    complete recorder, or deck and electronics?, 42-43
    how many heads ... motors?, 32-33
    judging quality, 37, 42
    other controls, 37
    price, 26-27
    push buttons or levers?, 33, 36
    reel to reel, cartridge, or cassette?, 27-29, 32
    supply sources, 43, 47-50
    what system?, 27
  portable, 74-79
    examples of, 76-78
  transport mechanism, 7-11
  typical stereo, 30-31
  ways of using, 70-72
Time indication
  audible, 94-96
  visual, 95-96
Time switch, use of, 80-82
Tone control, 63
Transistor microphone amplifier, 102-105

U

Ultrasonic bias. *See* High-frequency bias
Ultrasonic erase. *See* High-frequency erase
Unidirectional microphones, 56-58
Using a recorder
  examples of ways, 70-72
  learning a language, 87
  level setting, 59, 62-63
  party games, 83-84
  self-criticism, 82-83
  with a growing family, 69, 73
Using a time switch, 80-82

V

Variable reluctance microphones. *See* Magnetic microphones
Visual time indication, 95-96
Volume indicator, 62

W

Ways of using a tape recorder, examples of, 70-72
What system to buy?, 27
Wow, 8